THE LIFE OF
Sir ERNEST SHACKLETON

BOOKS BY SIR ERNEST SHACKLETON

THE HEART OF THE ANTARCTIC Being the Story of the British Antarctic Expedition, 1907–1909.

SOUTH. The Story of Shackleton's Last Expedition, 1914–1917.

LONDON WILLIAM HEINEMANN LTD

Sir Ernest Shackleton, aged 40, with his Younger Son, aged 3.

THE LIFE OF
SIR ERNEST SHACKLETON
C.V.O., O.B.E. (*Mil.*), LL.D.

BY

HUGH ROBERT MILL

> Do your best, whether winning or losing it,
> If you choose to play !—is my principle.
> Let a man contend to the uttermost
> For his life's set prize, be it what it will!
>
> ROBERT BROWNING

LONDON : WILLIAM HEINEMANN LTD.
MCMXXIII

G
875
S5M5

944318

MADE AND PRINTED IN GREAT BRITAIN

TO THOSE
WHO KNEW HIM BEST
BECAUSE
THEY LOVED HIM MOST

PREFACE

THE life of a man of action may fitly be presented as a continuous narrative of his doings, from which a reader may gain a clear sense of the personality and trace the growth of character.

Shackleton's life as it ran on seems in retrospect to have passed through three periods now presented in three books: the first, Equipment for the Achievement of the second; the third, Bafflement, which an unconquerable optimism saved from defeat. Each of the three periods includes a series of distinct but consequent experiences forming natural chapters.

Such an arrangement leaves little room for generalities, and requires an Epilogue in which the essence of the life may be sublimed from the facts.

With this plan in view I have tried to set out the life and strife of my old friend as a sort of moving picture, the scale of which varies in its different parts according to the importance of each to the life as a whole. The endeavour necessarily falls short of the ideal inspiring it; but I have done my best to chronicle a life, and a whole life, so far as this is possible, while observing such restraint and reticence as are demanded by regard for the feelings of others.

I have had free access to all available records, including diaries and intimate correspondence, and I have consulted Lady Shackleton at every stage of the work, in which she has afforded me the most ungrudging help. This may thus be regarded as an authoritative and responsible biography. It is right to add that my discretion has been unfettered both as to the facts dealt with and the manner of presenting them.

When I first began to know Shackleton well, on board the *Discovery*, we were drawn together by a common love of the poetry which had so large a part in shaping his life. Now I find that this bond helps me to survey his career from a view-point that may almost be considered his own, and gives me courage for the effort to make worthy use of the wealth of material which Lady Shackleton's unreserved confidence and the voluntary help of many friends have placed in my hands.

As a student of the history of polar discovery and as a geographer in personal touch with all the explorers of the last thirty-five years, I have been able to appreciate the work of Shackleton on his expeditions and to describe it with confidence and freedom; but I have had little to do with the business world, and the ways of the City are mysterious to me, so that I could not enter with like sympathy and completeness into the commercial enterprises which Shackleton pursued from time to time with all the impetuous ardour of his nature. Nor have I been able to follow him in his social life and relaxations when at home.

I have satisfied myself as to the substantial accuracy of every statement of fact and of all quotations. Most of the chapter-mottoes were favourites with Shackleton, and they were not chosen at random.

There have been some difficulties as to dates, for both Shackleton and some of his correspondents often contented themselves with the day of the week, and when envelopes were preserved the post-marks were not always legible; but the date of no important event referred to remains in doubt.

The material for the first three chapters of Book I. I owe almost entirely to Sir Ernest Shackleton's mother and sisters, who lent me many early letters and confirmed doubtful points from their private diaries. I have, in particular, to thank

PREFACE

Miss Shackleton, Miss A. V. Shackleton, whose help was exceptionally important, and Miss Eleanor Shackleton for this information. The dates and ports of call of the early voyages were not clear from the letters, and for settling them I have to acknowledge the ready help of the officials of the Marine Department of the Board of Trade and of the Committee of Lloyds.

Every chapter contains the names of friends who gave information which they alone could supply, or appreciations of Shackleton's character which they had special opportunities of observing : to all of these contributors, and to a few who prefer to remain unnamed, I wish to convey my heartiest thanks, and those of Lady Shackleton, through whom many of them were received. If to some I must also offer apologies for having condensed or curtailed what they wrote, I feel sure that they will take into account my difficulties in dealing with a large subject under a space-limit.

Finally, I must thank my wife, whose unfailing help alone made it possible for me to undertake this work or to bring it to completion.

H. R. M.

16th December 1922.

CONTENTS

PREFACE vii

BOOK I. EQUIPMENT

CHAPTER I. THE FAMILY AND THE FUTURE FIELD

February 15, 1874—The Yorkshire Shackletons—The Shackletons of Ballitore Abraham, Richard the Friend of Edmund Burke, Abraham the Second—The Shackleton Arms—Ebenezer Shackleton of Moone—Henry Shackleton the Father, and Henrietta Gavan the Mother of the Explorer—The Irish Descent through the Families of Gavan, Cary, and Fitzmaurice—Cook's Antarctic Voyage, 1772-1774—The Voyages of Bellingshausen, 1820; Weddell, 1823; Biscoe, 1831; Balleny, 1839, Ross, 1840-1843; and of H M.S. *Challenger*, February 1874 . . 3

CHAPTER II. EARLY LIFE. 1874-1890

Childhood at Kilkea—His First Penguin—Boyhood in Dublin—The Antarctic Tunnel—Love of Funerals—Youth in Sydenham—Preparatory School and Dulwich College—School Friends—Poor Scholars· Great Truants—Trying to sign on—Success of the Last Term—Ready for Sea . . . 17

CHAPTER III. THE *HOGHTON TOWER*. 1890-1894

First Voyage round the Horn to Iquique with rating of Boy—Apprenticeship to North-Western Company—Second Voyage to Iquique—Third Voyage round the Cape to India, thence to Mauritius, Australia, and Chile—Storms and Narrow Escapes—An Adventure at Tocapilla—His Love of Letters—Passes as Second Mate—The Power of the Stars 30

CHAPTER IV. IN THE SHIRES AND CASTLES. 1894-1901

He astonishes a Ship-owner—Sails as Third Mate of *Monmouthshire* to the East—Attempts at Verse—Passes as First Mate—Sails as

xii   CONTENTS

PAGE

Second Mate of *Flintshire* to Japan and California—Falls in Love
—Passes as Master Mariner at Singapore—Begins to read
Browning—Wreck of the *Flintshire*—Joins Union Castle Line
—Sails on *Tantallon Castle* as Fourth Officer—On *Tintagel
Castle* as Third Officer—His First Book *O.H M.S.*—Meets
Rudyard Kipling—Becomes F.R.G S —The *Carisbrooke Castle*
—Appointed on National Antarctic Expedition .   .   45

CHAPTER V   THE *DISCOVERY*. 1901-1903

Engaged to Miss Dorman—Joins R.N.R —Sails as Third Officer
of *Discovery*—Winters in M'Murdo Sound—Edits *South Polar
Times*—Accompanies Scott to Farthest South—His Diary of
the Journey—Breakdown from Scurvy—Invalided Home and
arrives well—Determination to Go Back.   .   .   57

CHAPTER VI. SHORE JOBS. 1903-1906

In London as Sub-editor of *Royal Magazine*—First Public Lecture
—Marriage—In Edinburgh as Secretary of Royal Scottish
Geographical Society—Electrifying a Council—Takes to Golf
—A Social Success—Birth of Son—Stands for Parliament at
Dundee—The Hecklers—State of the Poll—Resigns Secretary-
ship—Joins Beardmore Firm at Glasgow—Commercial Ventures
and their Result—Appreciation by Friends   .   81

BOOK II.   ACHIEVEMENT

CHAPTER I. SHACKLETON ASPIRES. 1906-1907

Plans an Expedition to the South Pole—Birth of Daughter—Tries
to enlist Old Comrades—Learns of Scott's Plans—The Promise
as to Land Base — Change of Plans — Raises Funds by
Guarantees—Buys the *Nimrod*—Completes Staff—Novel Ideas
for Stores and Transport—King Edward inspects Ship—Receives
M.V.O.—Travels to New Zealand and joins *Nimrod* .   . 103

CHAPTER II   THE *NIMROD*. 1908

Towed by *Koonya* to Antarctic Circle—Reaches the Ice Barrier
and finds Balloon Bight gone—Effort to reach King Edward
Land—The Battle with the Ice—The Battle in his Mind—
Forced to break a Promise—Lands at Cape Royds—Difficulties
overcome—The Ship departs   114

CONTENTS

CHAPTER III. SHACKLETON ATTAINS. 1908-1909

Autumn Sledging—Ascent of Mt Erebus—Wintering at Cape Royds—Editing and Printing *Aurora Australis*—The Motor-car—The Kinematograph—The Ponies—Spring Depot Journey—Selecting Companions for the Pole—Start with Three Men and Four Ponies—Hut Point—The Barrier Surface—Mount Hope—Discovery of Beardmore Glacier—Loss of Ponies—The Ascent to the Plateau—The Last Outward March—88° 26' S.—The Great Race with Death—Starvation and Dysentery—Finding Last Depot through Mirage—Race for the Ship—Back to New Zealand 125

CHAPTER IV. POPULARITY. 1909-1910

Reception in New Zealand, Australia, and Italy—Arrival in London—The Lion of the Season—Receives C V O.—Honours and Dinners—Cowes Week—Parliament grants £20,000—Stay at Balmoral—Lectures to the Geographical Societies of Europe—Received by Kings and Emperors—Gold Medals and Illustrious Orders—Knighted by King Edward—Popular Lecture Tours at Home and on the Continent—Consideration to Friends—American Tour—Triumphs and Troubles . . . 156

BOOK III. BAFFLEMENT

CHAPTER I. UNREST. 1911-1913

The Explorer as Business Man—Small Success in the City—Sighing for the South—Helping Filchner in his Expedition—Receives the News of Amundsen at the South Pole—The *Titanic* Disaster—News of Scott's Triumph and Death—Business Visits to New York—Decides to start a New Expedition. . . 183

CHAPTER II. THE *ENDURANCE* 1913-1915

Floating the Imperial Trans-Antarctic Expedition—The Search for Funds—Sir James Caird's Generosity—Purchase of the *Aurora* and the *Endurance*—Getting his Men together—Testing New Gear on Norwegian Glaciers—The *Endurance* ready to Sail—Outbreak of War—Admiralty refuses offer of Ships and Men—Mental Struggle—The *Endurance* sails—Delay at South Georgia—Penetrates Weddell Sea—Discovers Caird Coast—The Ship beset and drifting North—The Ship sunk—Unceasing Vigil 193

CONTENTS

CHAPTER III. THE *JAMES CAIRD* 1915-1916

Life on the Drifting Ice—Attempt to reach Land—The Floe breaks up—Takes to the Boats—"Old Cautious"—The Worst Hardships—All Hands reach Elephant Island—Desperate Counsels—Preparing the *James Caird*—Leaves Twenty-two Men under Wild—Sails with Five Men in the *James Caird*—Reaches South Georgia—Crosses unknown Mountains—Reaches Norwegian Whaling Station—When was the War over? . . . 213

CHAPTER IV. ALL'S WELL! 1916-1917

Tries to reach Elephant Island in *Southern Sky*—At the Falklands sends News Home—Second Attempt to reach Elephant Island in *Instituto de Pesca* fails—Third Attempt in *Emma* fails—A Letter to his Little Girl—The *Discovery* coming out—Fourth Attempt in *Yelcho* saves Wild and "the Boys"—Triumphal Reception in Chile—Learns that he is not wanted to save Ross Sea Party—Travels to New Zealand—Accepts Hard Terms—Reaches M'Murdo Sound as Subordinate on his own Ship—Beards a Committee—Hastens Home for War Work . . 229

CHAPTER V PRO PATRIA. 1917-1919

Recruiting Speeches in Australia and America—Home again—A Difficult Man to fit in—Mission to South America under Department of Information—Successful Propaganda—Returns in a Troop Ship—A Major's Commission—Equips Expedition for Spitsbergen—Recalled to join North Russian Expedition—At Murmansk—Brief Winter Visit Home—The Arctic Winter—Archangel—Resigns Commission . . . 248

CHAPTER VI. THE LAST *QUEST*. 1920-1922

Business Life again—Restlessness—Plans for a Last Expedition—Project for exploring North Polar Area—Buys the *Quest*—Resolves to explore Oceanic Islands and Enderby Quadrant of Antarctic—Financed by Mr Rowett—Equipping Shackleton-Rowett Expedition—The Boys rally to the Boss—Voyage of the *Quest*—Headwinds and Strain—Long Stay at Rio—Reaches South Georgia—Crossing the Bar 265

EPILOGUE 285

APPENDIX List of Distinctions, compiled by Lady Shackleton . 293

INDEX 295

LIST OF ILLUSTRATIONS

Sir Ernest Shackleton, aged 40, with his Younger Son, aged 3	*Frontispiece*
	FACING PAGE
Ernest Henry Shackleton, aged 11	23
Ernest Henry Shackleton, aged 16	29
The Full-rigged Ship *Hoghton Tower*	39
Sub-Lieut. E. H. Shackleton, R N.R , aged 27	56
Ernest Henry Shackleton, M.V.O., aged 33	103
The *Nimrod* in Ross Sea	116
The Hut at Cape Royds	126
Camp on the Barrier, with Ponies	135
Camp on the Beardmore Glacier	139
Farthest South, 1909	144
Back from the Farthest South	151
Lady Shackleton and the Children, 1914	186
Sir Ernest Shackleton, aged 40, in Sledging Dress	194
The *Endurance* beset in Weddell Sea	209
The *James Caird* approaching South Georgia	226
Major Sir Ernest Shackleton, C.V.O., aged 44	256
Sir Ernest Shackleton, aged 47	267
The Quest, unloaded	272
Memorial Cairn at Grytviken	282

MAPS IN THE TEXT

	PAGE
Antarctic Regions as known in 1874	11
Routes of the British Antarctic Expedition, 1907	124
Antarctic Regions as known in 1922	196
Routes of *Endurance*, *James Caird*, and Relief Ships, 1914-1916	236

BOOK I
EQUIPMENT

"Long, long since, undower'd yet, our spirit
Roam'd ere birth, the treasuries of God;
Saw the gifts, the powers it might inherit,
Ask'd an outfit for its earthly road.

Then, as now, this tremulous, eager being
Strain'd and long'd and grasp'd each gift it saw,
Then, as now, a Power beyond our seeing
Stayed us back and gave our choice the law."
MATTHEW ARNOLD

CHAPTER I

THE FAMILY AND THE FUTURE FIELD

> . . . " No family
> Ere rigg'd a soul for heaven's discovery
> With whom more venturers might boldly dare
> Venture their stake with him in joy to share."
>
> DONNE.

ERNEST HENRY SHACKLETON was born at Kilkea House in the barony of Kilkea, near Athy, Co. Kildare, on 15th February 1874. At the same time, on the outer edge of the most distant ocean, H.M.S. *Challenger* with all her men of science was lying amongst the icebergs close to the Antarctic Circle, which she was the first to reach by the power of steam. The wanderings of a ship on the sea are so far like those of a planet in the sky, that one might cast a happy horoscope for an explorer from this aspect of challenge to the South. Twenty-seven years had to come before the line of life of the newborn was to be interwoven with the parallels of highest southern latitude; but the soul of the explorer had already been "rigged for discovery" by a crowd of ancestors, and the field of his future fame had been spied out and made ready by gallant pioneers through many generations. A backward glance in the two directions may serve to trace the converging lines of ancestry and exploration to their meeting-point.

Clever Mendelians may some day be able to describe and explain a man's character from a study of the ascending ramifications of his ancestry. This is beyond our power; but we can at least hope to detect, in the full records of many of Ernest Shackleton's forbears, fragments of character which, as the kaleidoscope of his life revolved, grouped themselves into the new and surprising patterns of his mind. It may be that we shall find in the

4 THE LIFE OF SIR ERNEST SHACKLETON

blending of like qualities and in the juxtaposition of incompatibles, a key to the originality, quick wit and unexpected decisions which in time fitted the infant of Kilkea to solve problems of the far South to which the *Challenger* was pointing at his birth.

Shackletons, taking their name from the village of Shackleton, near Heptonstall, in the West Riding of Yorkshire, have been traced back to the thirteenth century, when they held the position of foresters under the de Warrens, and references appear to "the fighting Shackletons" in stories of the wars of the Border. A branch of the family settled in Keighley near the end of the fifteenth century, and Richard Shackleton of Keighley, a bowman, followed Lord Clifford to the battle of Flodden in 1513.

In another part of the West Riding there were Shackletons of some repute during the sixteenth century, one of whom, Henry Shackleton of Darrington, was married in 1588 to a daughter of the Rev. Anthony Frobisher, vicar of the parish, who was a brother of the father of Sir Martin Frobisher the great Arctic explorer. The relation of the Shackletons of Keighley to those of Darrington we have not been able to ascertain ; but it is interesting to contemplate the possibility of Sir Ernest Shackleton having had even remote ancestors in common with Sir Martin Frobisher

In 1591 one of the Keighley Shackletons purchased property in Bingley, where Shackleton House was inherited in 1654 by Roger Shackleton, who was born in 1616, a younger son of John Shackleton, and from him the descent to Sir Ernest Shackleton is clear. In 1675 Roger Shackleton settled the property on his son Richard, who was a man of strong religious convictions, and the first of his family to embrace the doctrines of the Society of Friends of Truth as preached by George Fox, the founder of " the people called Quakers." In the seventeenth century only those of high moral courage dared to be dissenters, and Richard Shackleton passed through no little persecution. He was imprisoned for three years in York Castle, and heavily fined for holding meetings of the Friends in his house. When the toleration of dissent was secured, his life flowed smoothly, and in 1696 he obtained a licence for his house as a place of worship·

In the same year his youngest son, Abraham, was born at Shackleton House. The house was the property of Shackletons until early in the nineteenth century; but it no longer exists, having been cleared away in 1892 in the course of street improvements.

Abraham Shackleton was a delicate child, and both parents died before he was ten. His education was so far neglected, that he was twenty years of age before he began to learn Latin. Realizing the seriousness of the case when, after trying various other occupations, he found that his only opening in life seemed to be that of the teaching profession, he made a supreme effort, and in a very short time obtained a mastery of the language, and, his granddaughter records, he was eventually able to write " pure and elegant Latin." After preliminary experience as a teacher in Skipton he was invited by some Quaker friends in Ireland to go there as a private tutor. He returned to England in 1725 to marry Margaret Wilkinson, a cousin of his Skipton headmaster, and the pair settled at Ballitore, Co. Kildare, thirty miles south of Dublin, where they opened a boarding school for boys in 1726. Although they were strict and uncompromising Quakers, the Shackletons won the respect, confidence and affection of people of all creeds. What wise indulgence to youth Abraham Shackleton's Quaker principles permitted, and how he attracted the confidence of his boys, may be judged by this extract from a rhymed epistle of nine couplets written by his little son, Richard, then aged eight.

"DUBLIN : *The 10th of the 5th mo.*, 1734.
"HONOURED FATHER

Since my last I've seen the Fair
And many Tents and Drunkards there.
Six foreign Beasts I went to see
And Birds of which one frightened me.
I've seen at last the mighty Sea
And many ships near to the Quay.

. . . . .

I've seen the College and the Castle
And many boys that love to wrestle.
With dearest love I now conclude
And always hope I may be good.
RICHARD SHACKLETON."

6   THE LIFE OF SIR ERNEST SHACKLETON

Some famous men were educated at Ballitore, which in time came to be known as "The Eton of Ireland." Richard Brocklesby, the friend and physician of Dr. Johnson, was one of these. Another was Edmund Burke, who venerated Abraham Shackleton throughout his life, believing that from him came all the good he ever got from education, though he spent but two years at Ballitore.

Burke was twelve years of age when he went to the Quaker school, and made the acquaintance of Richard Shackleton, the son of the headmaster, who was three years his senior. Between them, in the words of Viscount Morley, " there sprang up a close and affectionate friendship, and, unlike so many of the exquisite attachments of youth, this was not choked by the dust of life, nor parted by divergence of pursuit. Richard Shackleton was endowed with a grave, pure and tranquil nature, constant and austere, yet not without those gentle elements that often redeem the drier qualities of his religious persuasion. When Burke had become one of the most famous men in Europe, no visitor to his house was more welcome than the friend with whom long years before he had tried poetic flights, and exchanged all the sanguine confidences of boyhood. And we are touched to think of the simple-minded guest secretly praying, in the solitude of his room in the fine house at Beaconsfield, that the way of his anxious and overburdened host might be guided by a divine hand."

Richard Shackleton studied at Trinity College, Dublin, and was the first Quaker to do so, although he could not graduate, being a nonconformist. He married in 1749 Elizabeth, daughter of Henry Fuller of Fuller's Court, Ballitore, a house with a beautiful garden, famous for its clipped yew hedges, and laid out on the rich bog-land reclaimed by the settlement of English Quakers at the end of the seventeenth century. His son Abraham was born in 1752, and his wife died two years later. He married again in 1755, Elizabeth Carleton, by whom he had a daughter, Mary, who, as Mrs. Leadbeater, was famous as a graceful writer, and in the *Ballitore Papers* gave the best account of social life in Ireland during the great rebellion. In her *Memoirs and Letters of Richard and Elizabeth Shackleton*, pub-

THE FAMILY AND THE FUTURE FIELD

lished in 1823, she says of her father : " The fault of his temper was quickness, not violence ; but this was soon subjected to his judgment, and if he thought he had wounded anyone thereby, he was ready to acknowledge it with a benign humility."

Richard had succeeded his father as headmaster of Ballitore School in 1756, and in 1779 he retired, handing on the succession to his son Abraham. From an early age until shortly before his death in 1792, Richard Shackleton had represented the Irish Quakers at the Friends' Yearly Meeting in London, and at Burke's hospitable table he met the great men of the day, and shared in the brilliant conversation of the London of Dr. Johnson. There can be no doubt that he was familiar with the voyages of Captain Cook, and knew how the myth of a vast Southern Continent was shattered by Cook's voyage in H.M.S. *Resolution* in 1772-75, when the Antarctic Circle was first reached and found to girdle a region of perpetual ice. But he could not know that his own great-great-grandson was destined to link the name of Shackleton for ever with the solitary snow-clad Isle of Georgia which Cook had added to the British realm.

The younger Abraham Shackleton was a student of natural science as well as of classical learning, and wrote memoirs on astronomy, botany and conchology based largely on his own observations. He shared the tastes of his father and grandfather in literature and poetry, delighting in Milton and Cowley. He also followed the example of his grandfather Abraham by coming to England for his wife. He married in 1779 Lydia, daughter of Ebenezer Mellor of Manchester. Her mother was a daughter of John Abraham of Swarthmoor Hall, Ulverston, and a direct descendant of Margaret, wife of Judge Fell of Swarthmoor Hall and later wife of George Fox the founder of the Society of Friends, a woman of heroic courage and unconquerable will. She wrote four outspoken letters to Oliver Cromwell on the persecution of the Quakers, and after the Restoration she journeyed to London and spoke plainly to Charles II. of the continued persecution of his most loyal subjects. The mind of Abraham Shackleton was turned to the history of his family, and in 1794 he had a correspondence with a cousin then residing in London, who conducted inquiries in

8   THE LIFE OF SIR ERNEST SHACKLETON

Lancashire and not only made out the genealogy, but forwarded to the Quaker schoolmaster a copy of the coat-of-arms of Richard Shackleton in 1600. These arms, as subsequently confirmed to him by the Herald's Office in Dublin, showed three gold buckles on a red ground, and bore the crest of a poplar tree with the inspiring motto, *Fortitudine Vincimus.*

Ebenezer Shackleton, the son of Abraham and Lydia, born in 1784, was thus descended on both sides from the sturdy stock of original North of England Quakers. He became the owner of flour mills at Moone, Co. Kildare, and was an enthusiastic horticulturist. The garden surrounding the beautiful house which he built at Moone was laid out with great skill and kept in proud perfection. His political principles were said to be far in advance of his time, and he was a close friend of Daniel O'Connell the patriot, and of Father Mathew the famous temperance reformer. He was twice married, his second wife being Ellen, daughter of Captain William Bell of Bellview, Abbeyleix. She was a woman of rare charm, both in person and character, of cultured literary tastes, and came of Quaker stock. Ebenezer Shackleton was formally removed from the Society of Friends for his failure to conform strictly to its rules in some particulars; but he continued to attend the meetings, and on his death he was laid in the Quaker burying-ground. The children of the second marriage were brought up by Mrs. Shackleton as members of the Church of England. Henry Shackleton, destined to become the father of the explorer, was one of the younger sons of Ebenezer and Ellen Shackleton. He was ten years old at his father's death, and his mother hoped that he would adopt the army as a profession, so he was sent to the Old Hall School in Wellington, Shropshire. The boy's warlike reputation may be gathered from the description of a snowball fight in the school magazine:

> " Oh never did so brave a band
> With such a threat'ning aspect stand
> Upon those rocks before!
> The savage hordes of Shackleton
> The Langtry and McLaine
> And all the lowland chivalry
> Were gathered on the plain "

A serious illness destroyed his chance of going in for the artillery as he had intended, and on recovering he was sent to a school at Kingstown. Here he won an exhibition to Trinity College, Dublin, where he graduated in Arts in 1868, winning the silver medal. From the earliest time the family had loved the country, and the care of farm and garden was natural to them. The three generations at Ballitore had cultivated their land as well as the minds of their scholars, and almost instinctively Mr. Henry Shackleton bought a small property at Kilkea, Co. Kildare, and devoted himself to farming with all his heart. In 1872 he married Miss Henrietta Letitia Sophia Gavan, the only daughter of Mr. Henry John Gavan, son of the Rev. John Gavan, Rector of Wallstown, Cork.

Miss Gavan had long known Mrs. Ebenezer Shackleton, for whom she had an extraordinary affection, and as a girl she used to say that she would only marry a son of Mrs. Shackleton, though at the time she did not know her future husband. Miss Gavan was of Irish descent, and this marriage was the first infusion of Celtic blood into the sturdy Quaker stock which had remained English though living in Ireland for a century and a half.

The Rev. John Gavan lived in troublous times, and many tales were told to his great-grandchildren of his adventures during the Whiteboy riots of a hundred years ago His daughters shared his personal courage, and when for many months the rectory was garrisoned by troops for their protection, the girls were taught to shoot by the soldiers. They were strictly interned within the boundary of the glebe, but one day they determined to break bounds and went for a drive. They were soon stopped by a party of riotous turf-cutters, who brandished their long spades and barred the way. The officer of the little garrison was coming up cautiously to reconnoitre when his horse, which belonged to one of the young ladies, hearing his mistress's voice, broke into a gallop and dashed into the thick of the rioters, scattering them in terror. Several members of the Gavan family went out to Tasmania and New South Wales in the early days of settlement.

Mrs. H. J. Gavan, the mother of Mrs. Henry Shackleton, was a daughter of John Fitzmaurice and Henrietta Frances Cary.

The Fitzmaurice strain may be traced back to the 20th Baron of Kerry, and the Irish annals contain many records of the reckless daring of "the turbulent Fitzmaurices." One of her first cousins was a commander in Nelson's flagship at the battle of Copenhagen. Her grandfather on the Fitzmaurice side used to boast that he had run through estates in nine counties, and the lavish extravagance of his living was illustrated by his habit of regaling his huntsmen with mulled port

Miss Henrietta Cary's grandfather was the Archdeacon of Killala, whose nephew was the Rev. Henry Francis Cary, the translator of Dante. On the Cary side there was a Huguenot ancestor, a refugee from France in 1685, who was also an ancestor of the great polar explorer, Sir Leopold McClintock.

Mr. H. J. Gavan, father of Mrs. Henry Shackleton, qualified as a medical man, but deserted medicine for a commission in the Royal Irish Constabulary, and shortly after his marriage he received the important appointment of Inspector-General of Police in Ceylon. He and his wife embarked at Falmouth in 1844 in the sailing ship *Persia* for the long voyage round the Cape of Good Hope, but the start was inauspicious. The ship met a fierce gale in the Channel, and after struggling against it for ten days had to put back to Falmouth. Here several of the passengers, who had suffered so badly from sea-sickness that they could not resume the voyage, were put on shore. Amongst these were Mr. and Mrs Gavan, who had to leave all their heavy luggage behind in the ship's hold A small brown wooden chest, part of the cabin equipment, came ashore and was a familiar relic in the nursery of their grandchildren. Mr. Gavan never recovered from his cruel tossing on the sea, and died in 1846 soon after the birth of his daughter. Mrs. Gavan long survived to be a beloved grandmother to her daughter's family.

Courage, cautious in the English line, reckless in the Irish, idealism (chastened on the one side, fantastic on the other), devotion to great religious enthusiasms, love of free life in the open air, a passion for poetry, and an unfailing instinct for friendship, were common to both the Shackleton and the Gavan descent When Ernest Henry was born on 15th February 1874,

THE FAMILY AND THE FUTURE FIELD 11

the second child of Mr. and Mrs. Henry Shackleton, he brought with him a great inheritance of splendid qualities. To what use he put it will appear as we proceed to map the torrent of his impetuous life.

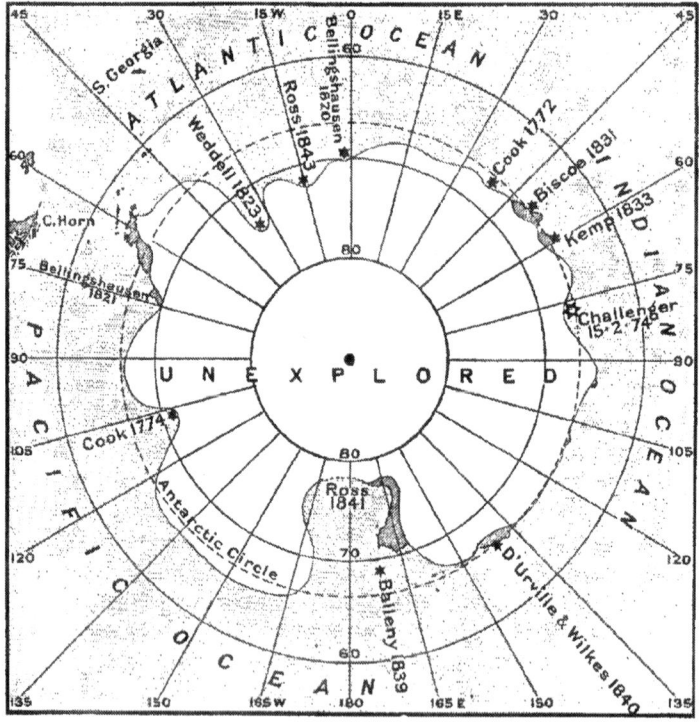

ANTARCTIC REGIONS AS KNOWN AT SHACKLETON'S BIRTH, 1874.

The other event of 15th February 1874, the arrival of H.M.S. *Challenger* at the Antarctic Circle, requires for its full appreciation a backward glance along the history of discovery by sea. Captain Cook died convinced that he had settled the problem of the Antarctic continent, so that no explorer would henceforth care to follow it within the belt of ice. He had, however, supplied a magnet that drew a fleet of sealers to southern waters by his discovery of the Isle of Georgia, a name which, apparently by the carelessness of an engraver, was changed to South Georgia. American independence had diverted much of the South American sealing trade to British ports, and the firm of Enderby Brothers, in particular, sent their small vessels from the Thames to South Georgia and other sub-Antarctic islands. The close of the Napoleonic wars in 1815 threw many naval officers out of employment, and some of the most adventurous of them took to Antarctic sealing; and being good navigators and provided with chronometers, they were able to make charts of their discoveries and give a clear account of them.

In 1819 a British trader between the River Plate and Valparaiso discovered a group of snow-clad islands far south of Cape Horn and called them the South Shetlands. The most easterly islands of the group were named Elephant Island, from the abundance of sea elephants, and Clarence Island. In the same year the Tsar Alexander I sent out two great expeditions, one to the North Pole, the other in two ships, the *Mirni* and *Vostok*, to the South Pole. To Bellingshausen, the commander of the southern expedition, was assigned the task of supplementing Cook's Antarctic voyage by going south where Cook had kept to the north, and keeping north where Cook had gone south. He made a complete circumnavigation, and in particular sailed far within the Antarctic Circle along a high ice-wall sometimes reaching 69° S. in what was subsequently called the Enderby Quadrant, *i.e.*, the range of longitude from 0° to 90° E.

In 1823 James Weddell, a Scottish sealer and ex-naval officer, with two small vessels, the brig *Jane* of 160 tons, and the cutter *Beaufoy* of 65 tons, found the ice conditions to the south of South Georgia exceptionally favourable, and succeeded in reaching to 74° S in 34° W., the farthest south hitherto attained, and he reported open sea there. He called the part of the ocean he had traversed King George IV. Sea; but posterity has honoured the discoverer by naming it the Weddell Sea. No subsequent explorer has found this sea clear of drifting floes, and much will be heard of it in later chapters. An entertaining American sealer and author, one Benjamin Morrell, described a voyage in the same sea in the same year, in the course of which he caught seals on the coast of New South Greenland, which is still sometimes found dotted-in on Antarctic maps under the name of Morrell Land. It was said to lie in longitude 50° W., and much controversy has been wasted upon it. It was probably only the east Graham Land coast misplaced by the rough-and-ready calculations, or by the crafty design of a sealer bent on his trade. Morrell Land will beckon to us again amid the mists.

Ten years elapsed before the next discoveries. Then in 1831–33 John Biscoe, an ex-naval officer with two of Enderby's ships, the little brig *Tula* and the cutter *Lively*, made a magnificent circumnavigation in the far south, following Bellingshausen's track through the Enderby Quadrant ; though he had never heard of the Russian explorer, whose work remained sealed up in the Russian language for many years. He had the good fortune to discover land on the An'arctic Circle in 50° E. He saw it in bad weather and made heroic efforts to reach a bold headland, which he named Cape Ann, fighting against incessant gales with a crew reduced to a few men by illness and hardship. It is the Enderby Land which remains on the map, though no human eye has seen it since its discovery. After refitting at Hobart, Biscoe resumed his voyage eastward and discovered the west coast of Graham Land and the Biscoe Islands south of the South Shetlands. On his return to England he found that the Royal Geographical Society had been established, and he received from it the first gold medal ever given for Antarctic exploration. He was a fine navigator, a keen observer, and his observations on sea-ice anticipated many conclusions arrived at by better equipped expeditions long after his work was forgotten. A second expedition was prepared for him with the aid of the Admiralty, but it led only to disputes and disaster.

Another of Enderby's sealers, John Balleny, in the little schooner *Eliza Scott* with the cutter *Sabrina* in company, left Campbell Island, south of New Zealand, in January 1839; and keeping farther south than any one else had done in that region, he fell in with a group of mountainous islands, now known as the Balleny group, on the Antarctic Circle, in longitude 164° E.

14 THE LIFE OF SIR ERNEST SHACKLETON

It is impossible to praise too highly the courage and tenacity of the early sealers in their ill-found little vessels, especially when it is remembered that the time spent on discovery, and still more the publication of the position of new coasts, meant a reduction, sometimes the total loss, of the profits which might be expected from their trade.

The next impulse to exploration in the far south came from the side of science. The increase of shipping on the routes from the Atlantic to Australia and China made a better knowledge of the magnetic conditions of the Southern Ocean a matter of practical concern, and about 1835, expeditions, in which magnetic observations in those seas had an important place, were planned in France, the United States and Great Britain. In the two former the Antarctic episodes were but a small part of the programme, and it is sufficient here to note that both D'Urville in the French expedition and Wilkes in the American tried to enter the Weddell Sea and were forced by ice to abandon the attempt. In January 1840 both expeditions discovered land, called by the French Adélie Land, on the Antarctic Circle south of Tasmania, and Wilkes believed that he saw signs of a continent along 1500 miles of ice-laden sea to the west.

The British expedition was specially fitted for magnetic work in the South Polar area, and the command was given to Sir James Ross, who had already located and visited the North Magnetic Pole, and was ambitious of paralleling this exploit in the south. He was in command of a stout ship of 350 tons, H M S. *Erebus*, and Captain Crozier commanded H.M.S. *Terror*, a similar vessel, the two sailing in company. Naval officers were trained for magnetic work, and the surgeons were expected to deal with other branches of science. The youngest man on the expedition was Joseph Dalton Hooker, assistant surgeon on the *Erebus* and already a keen botanist. The expedition sailed in 1839 and returned in 1843, having spent three successive summers in Antarctic exploration. The first two seasons, 1840-41 and 1841-42, were devoted to the region south of New Zealand, where Ross found it possible to penetrate a broad zone of floating pack ice and entered a stretch of nearly open water, since called the Ross Sea. The western boundary of this sea

THE FAMILY AND THE FUTURE FIELD 15

Ross found to be a coast rising inland to a lofty range of mountains, several exceeding 10,000 feet in height, and stretching from a low promontory, Cape Adare, in 71° S. southward for 400 miles, where at M'Murdo Bay, nearly in 78° S., it turned sharply to the east. The coast was called Victoria Land, and the magnetic pole was located far in its interior, not to be reached by sea. No landing was made on the mainland, but a ceremony of annexation was held on Possession Island, which lay off the coast. The eastern side of M'Murdo Bay was rocky land, rising to two great volcanic peaks, one active, one extinct, to which with singular appropriateness the ships stood godfather and left Mounts Erebus and Terror to mark their farthest south. The land at its eastern extremity, Cape Crozier, joined on to a huge wall of ice apparently afloat—deep water reached to near its vertical face—and stretching for 400 miles to the eastward, with a height varying from 200 to 300 feet. At the eastern extremity of the Barrier, signs of land were found, but Ross was too cautious to claim this as a discovery. The marvel of the Ice Barrier impressed the explorers more than anything else, for it resembled nothing that they had ever seen, and its origin was a mystery. It was conjectured that the great flat-topped icebergs, or rather floating ice islands, of the Antarctic seas were portions of this or some similar barrier broken off by earthquakes or gales.

A great gap had been cut by these voyages into the disc of the unknown bounded by the Antarctic Circle.

The summer of 1842–43 was devoted to a strenuous attempt to follow Weddell's track into the Weddell Sea on the opposite side of the Antarctic area, the only other known gap in the disc. All attempts to penetrate the pack failed near the line of Weddell's track, but 19 degrees farther east a way was forced, and the *Erebus* and *Terror* attained the latitude of 71° S., and a sounding was made which Ross believed indicated the tremendous depth of 4000 fathoms.

In 1845 some additional magnetic observations were made by Moore on the *Pagoda* in the Enderby Quadrant, which Ross had not visited; but the attempt to get to Enderby Land was foiled by bad weather.

For nearly thirty years no ship troubled the solitude of Antarctic waters. A great expedition for studying the depth and the living creatures in all the oceans of the world was equipped in 1872 by the British Admiralty on H.M.S. *Challenger* under the command of Captain George Nares, R.N., with a scientific staff of six men, presided over by Professor Wyville Thomson of Edinburgh. One of the staff was John Murray, then unknown, later to become the most renowned oceanographer in the world. This expedition made a half-furtive dash across the Antarctic Circle south of Kerguelen Land on 16th February 1874. It was a hazardous adventure, for though the *Challenger* had the benefit of steam power, she was by far the largest ship that ever entered the southern ice, and she had no special strengthening to resist pressure if caught in the pack. In her short sojourn a number of soundings were made, samples of the deposits on the bottom of the ocean were secured, the icebergs were photographed for the first time, and material was collected by which subsequent laboratory investigations proved conclusively that somewhere not very far to the south of the *Challenger's* track there lay a great continent on which no human foot had trod. A grand result, full of the promise of new research, but again the mists rolled round the southern land, and for nearly twenty years the solitudes remained more menacing than inviting.

CHAPTER II

EARLY LIFE. 1874-1890

" Heaven lies about us in our infancy;
The soul that rises in us, our Life star,
Hath had elsewhere its setting and cometh from afar
. . . . . .
The Youth, who daily further from the east
Must travel, still is Nature's Priest,
And by the vision splendid
Is on his way attended "

WORDSWORTH.

WE search in vain amongst the memories of his mother, his sisters, and his school friends, above all in his school reports, for any intimation of future greatness in the childhood of Ernest Henry Shackleton. Many a man who jogged dully along life's highway for threescore years and ten has started as an infant prodigy or has at least shone resplendent in his school prize lists and first elevens. But now and again a Clive or a Darwin, who had been small credit to his teachers, has become a light to the world. So it was with Shackleton.

His was a natural and happy childhood; he had perfect health, a generous spice of mischief, and a lively fancy. He responded to the love of his parents, and as the years went on he accepted the devotion of the growing band of sisters as one of the native rights of man. His life until he went out into the world at sixteen falls into three parts—six years at Kilkea, four years in Dublin, and six years at Sydenham, always living at home.

Kilkea, in the most fertile part of Co. Kildare, is within five miles of Moone, the residence of Ernest's grandfather, and of Ballitore, made famous by the Shackleton family. Quiet rural beauty without much scenery save such as rich meadows, cultivated fields and woodlands can afford, is the feature of the district; and if the garden of Kilkea House was less perfect horticulturally than that at Moone, it was a pleasant place for children to grow up in. Mrs. Shackleton says that for the first eighteen months of Ernest's life she feared that he was too good to live. He never cried; his blue eyes and golden hair gave him an angelic beauty; but as the years passed, the alarming symptoms abated.

Mr. Shackleton was devoted to his garden, his farm and his children. He was their companion and instructor in the life of the open air. He taught them to swim in the river Griese, reminding them of the old Quaker joke of the Ballitore schoolboys, "Those who go into the Griese come out dripping." He set them on horseback at a very early age; he encouraged them to jump from high and ever higher places, to shun no reasonable risks, and to be afraid of nothing. Though courage, both physical and moral, ran in their blood, the little Shackletons were not exempt from fear, and an incident is remembered in which Ernest, frightened rather by the superstitious tales of a silly nursemaid than by the noise itself, asked to be shut up in the nursery cupboard while a great thunderstorm was raging. It is, however, worth noting that even after he became famous as the most fearless of explorers, he did not like this episode of his infancy to be referred to.

One might have expected to hear of him even as a child eager to be first in everything that was doing and in the front on all excursions. The reverse seems to have been the case, for in the later Kilkea days, when the little group of children went out with their nurse in the country lanes, Ernest was always dropping behind, looking about him in the hedges and ditches, happy in his own thoughts, or gathering and sometimes eating the wayside flowers. So habitual was this, that the nurse used to call him "Mr. Lag." It is almost certain that Ernest could read simple words when he was four years old. He was much attracted by music, and would listen in great contentment to his mother singing Irish airs, or playing on the piano for an hour at a time. When a child he took an extraordinary fancy to a feather muff belonging to one of his sisters and always wanted her to lend it to him for the pleasure of admiring and stroking it—an incident too trivial to mention save for the fact that the muff was made of the skin of a penguin and came from the far south.

For several years the Shackletons lived in Kilkea House, and there six children were born, two boys and four girls. The general depression of agriculture all over Ireland in the late 'seventies turned the mind of Mr. Shackleton towards a new profession, and as he had long been interested in homœopathy, of which his uncle, Dr. William Bell, was a practitioner, he decided at the age of thirty-three to start the systematic study of medicine. In 1880 he re-entered, at Trinity College, Dublin, the classrooms which he had quitted as a graduate in Arts twelve years before. At the same time he grew a beard, and this brought out so striking a resemblance to "the uncrowned king," whose sway in southern Ireland was then in the ascendant, that the lively youngsters in the medical classes dubbed him "Parnell" forthwith. The nickname was probably not altogether uncongenial, as he was and remained a strong believer in Home Rule for Ireland in the moderate form demanded in those reasonable days. The family settled at 35 Marlborough Road, and although two years younger than his great-great-grandfather was when writing the rhymed account of Dublin's wonders, quoted in the last chapter, Ernest must have looked with no less interest at "the mighty sea, and many ships near to the quay." His grandmother, Mrs. Gavan, about this time told how her brother, who was in the Royal Navy, was seized by Spanish pirates and was never heard of again; the children looked for the return of this hero of romance until the lapse of years showed it to be impossible. Stories of the sea no doubt gave point to his favourite game of "cabins," which led to his constructing a famous ship in the Dublin garden. The hull was a garden-frame which he decked over with boards detached from a garden seat, thus forming a dark hold, which he proceeded to stow with a diversity of cargo. Not ships, however, but funerals were his chief joy. It was almost impossible to keep him from running after a funeral if he sighted one anywhere in the streets. A possible explanation is that this was the easiest expression of a child's natural love for pageants, funerals being the commonest and least dangerous things of the kind to be met with in Dublin in the early 'eighties. The children were taught to shun brass bands, if not actually to take cover at the sound of one ; for a brass band was the common sign of a fenian or other rebel procession, and in the vicinity of such a challenge to the civil power there were risks which the little Shackletons' father did not consider reasonable.

One of the institutions of the Dublin days was an annual picnic to the Three Rock Mountain, an imposing mass south of the city, which soars, indeed, to something short of 1500 feet above sea-level, but has some rough ground and steep slopes on its sides. Here their father made the children, boys and girls together, run races downhill to a given mark and back again, his advice at the start being "stick your heels in and keep your heads up !" This early familiarity with steep places must have come back years later to the mind of the Antarctic leader, encouraging his men on mountains that surpassed the Three Rock even as the man had outgrown the child

The children had lessons from a governess, and one day the form of the Earth was the subject, and stress was laid on the fact that the lands of the Antipodes were situated directly beneath our feet. Ernest was then about seven years old, and after some meditation he enlisted all the available labour the nursery afforded, and set to work in the back garden to excavate a shaft by which he hoped that a new and shorter route to Australia could be established. The depth was already formidable to the small workers, when the threat of the landlord's displeasure enabled the domestic authorities to secure the abandonment of the scheme. So his first Antarctic expedition might be said to have ended, not from any fear in the leader, but on account of the unreasoning prejudices of people who could not understand It must be acknowledged that the real attraction of the tunnel was the delight of digging in the earth. At a time when most boys of his period looked forward to becoming postmen or engine-drivers, "And what are you going to be when you grow up, my little man ?" drew from him the solemn declaration, "a gravedigger." Stories of buried treasure fascinated the boy, and they never lost their glamour for the man, nor to the young Irish mind did it detract much from the joy of search that the treasure must be buried before it could be found, it only required some careful planning. One day Ernest confided to a favourite servant that he was of opinion that there was buried treasure in the garden, possibly gold and gems in dazzling profusion, and he was willing to suggest likely spots for the search if she would dig. The maid agreed, perhaps with a suspicion that the claim had been "salted," and sure enough treasure was found—a ruby ring belonging to Mrs Shackleton The episode seems to foreshadow the powers of organization and persuasion which were afterwards developed to such good purpose, but at the time it led only to a painful vindication of the rights of property.

Though no longer "too good to live," Ernest was an affectionate and lovable boy, his kindness to his little sisters was inexhaustible. He used to be fond, when seven or eight, of a book of Arctic travel—C. F. Hall's *Life with the Eskimo*, the pictures in which were an inexhaustible attraction, showing as they did ice-floes, towering bergs, snow houses, and the hunting of great beasts on land and sea. His first geography book also had pictures of Arctic and Antarctic bergs.

He early acquired a love for poetry, as who would not with his heredity ? His father revelled in Tennyson, but read largely in the other poets, and his mother matched him in her tastes. It was a usual thing at meal-times in the Shackleton household in Dublin and Sydenham, for the father to quote a verse and demand, " Where is that from ? " and there was some keenness of competition to be first with the correct reply. Between themselves the children played enthusiastically at capping verses, so their memories were stored betimes, and verses rose spontaneous to meet every call in later life. At an early age Ernest used to recite Macaulay's " Lays," and from the fervour of his declamation there could be no doubt that he felt himself every inch Horatius keeping the bridge. It was the same with other favourite pieces, such as "Casabianca," "The Burial of Sir John Moore," " The Burial of Moses," " Ye Mariners of England," and "The Wreck of the Hesperus." All these were learned in the Dublin days and never forgotten. The passion for realizing the heroic almost went beyond the limits of unconscious dramatic art. A few years later he actually made his younger sisters believe that the Monument near London Bridge was erected in honour of himself and his chum Maurice Sale-Barker, because they had put out a big fire which threatened to destroy the city. This no doubt was simply to test the credulity of their simple faith, but the significant thing is that even in early boyhood his soul yearned for great achievements and the honours they bring.

The Dublin days soon passed. In 1884 Dr. Shackleton took his medical degree at Trinity College after a distinguished course of study, and he also passed as a member of the Royal College of Surgeons of England In December 1884 Ernest Shackleton made his first voyage crossing the Irish sea in the *Banshee* with his brother and seven sisters Dr. Shackleton had now a family of nine, and the whole household moved to South Croydon in Surrey. His efforts to create a practice were rendered the harder by his adoption of homœopathy, whereby he fell out of sympathy with the majority of his professional brethren, and it was no more easy for a medical man in the reign of Victoria to make his way against professional conventions than it was for the Yorkshire Quakers under Charles II. to hold their own against the tyranny of conformity. But like them he kept to his principles and only worked the harder. After six months, Dr. Shackleton left Croydon and took up his abode in Aberdeen House, 12 West Hill, Sydenham, adjoining St. Bartholomew's Church, half a mile from the Crystal Palace. Here he built up a practice, and lived for thirty-two years, trusted and loved by his patients, and cultivating his garden, for he was always a keen horticulturist, and became a great authority on roses—sought after as a judge in flower-shows all over the country. He found time in his busy days for literary work also, especially the reviewing of medical and scientific books.

From the opposite side of the wide road it may be seen that the top of the steeply-sloping roof of Aberdeen House is flat.

Photo.] [P. Mitchell.

ERNEST HENRY SHACKLETON, AGED 11.

Face p. 23.

To this lofty and unrailed deck Ernest and the other children were not long in finding a way; and in their surreptitious antics there they earned many a thrill of real danger escaped by luck rather than circumspection. In the autumn of 1885 Ernest, who had hitherto shared the instruction given by the governess at home, took a great step in his career. He began to attend Fir Lodge Preparatory School, a few hundred yards from his home, where Miss Higgins presided with firmness and originality over her charges. On the first day one feature of the discipline delighted him, as he was not inclined to the fault on which it was exercised. Miss Higgins enforced the rule that every boy who gave way to tears should expiate his weakness by nursing a doll in full view of the class.

He was now a sturdy, broad-shouldered boy of nearly eleven, full of life and noise, with an accent that, despite the parental care which had refined it by comparison with the brogue of Dublin, struck the ears of Londoners as terribly Irish. Thrown suddenly amongst the smart little Cockneys, whose fathers were somethings or even somebodies in the City, the young Irishman would have been doomed to the nickname *Paddy* but he had been forestalled, there was a Paddy there already, so he had to answer to *Mike*, and Mike or Micky he remained to some of his dearest friends to the end of his days. *Mike*, a kindly but faintly contemptuous name, was partly superseded by the altogether respectful *Fighting Shackleton*, an unconscious echo of his father's Old Hall days and of the Border wars six centuries earlier. This proud title was conferred after an impetuous attack on a bully who was tormenting a much smaller boy.

Of his last year at Fir Lodge, Mr. S. G. Rhodes writes:

"He was a general favourite by reason of his rather quiet, sensible ways. He was always a chum of mine, but the only incident standing out clearly in my memory is the fact that our schoolfellows decided that on St. Patrick's Day Mike and I were to have a scrap in honour of the sacred memory of St. Patrick. This we had, much to the elation and amusement of the rest of the school, and I believe entirely to our own enjoyment."

A photograph taken in 1886 shows Ernest hauling on a rope in a photographer's setting of a ship's deck; a broad hint as to a secret inclination for the sea which had come upon him. That he already knew much of travel and exploration may be gathered from an interview published in *M.A.P.* in 1909, when he is reported as saying : " I have always been interested in Polar Exploration. I can date my first interest in the subject to the time when I was about ten. So great was my interest that I had read almost everything about North and South Polar Explorations "

In the Summer Term of 1887 Mike passed on to Dulwich College as a day boarder, and here he remained until the end of the Lent Term 1890, making remarkably little impression either on his masters or on most of his companions. A day boy, of course, has much less chance than a full boarder of being moulded by the traditions of a public school, and to one not over-well prepared by earlier education and living in so vivacious and unconventional a family as he did, the school life must have been somewhat uncongenial. He was certainly a poor scholar, always in a form the average age of which was about a year less than his own, and in the class lists his name was almost always far south of the equator and sometimes perilously near the pole. Only in his last term did he attain the position of fifth amongst thirty-one boys, his previous term having left him in the more usual place of twenty-fourth, amongst thirty. This proves that if there had been a strong enough incitement he could have done very well ; but somehow the masters failed to touch the spring which controlled his ambition. The earlier reports of his form-masters abound in such phrases as " he has not yet fully exerted himself," " wants waking up, is rather listless," " not a very clear-headed boy yet," " often sinks into idleness," " must remember the importance of accuracy " ; but they almost all go on to say that with increased attention he might do well, and that his abilities were good. The Headmaster found him " backward for his age." The boy's heart was anywhere but in his work until his last term. Even his schoolfellows remember him dimly, one as " doing very little work, but if there was a scrap on he was usually in it " Another, as "a cheery though not exuberant sort of fellow "

Mr. John Q Rowett says : " We used to walk to school

together very often, but I was never in the same form with him at Dulwich College, as he was three years older than me and was in a higher form. He was always full of life and jokes, but was never very fond of lessons, and I remember we had great difficulty with our modern and classical languages I was a terrible duffer at Greek, and he used to help me sometimes ; while I had a friend who knew German very well, and I used to get hints from him which I passed on to Shackleton "

The kindly help was of small avail, for the judicial summing up in the next report ran, " The results of the German examination were disastrous To set against this there has been excellent work in French "

He entered heartily into the school games, especially cricket, in which he was something of an enthusiast, always holding the game in high esteem for its value as training. Football, too, had its attractions ; and as he was a big boy for his age and always keen when his spirit was roused by exertion, he probably did well in it. He excelled in gymnastics, and delighted in feats on the trapeze All through his school years he was, like most boys, a collector of all sorts of things He was observant on his walks, and picked up fossils and minerals ; he never tired of hunting for " gold " in coal or slates, collecting the bright yellow cubes of iron pyrites as some approximation to hidden treasure. He would have responded heartily to instruction in Nature study, but there is no record of his receiving any. At home he was devoted to carpentry, his most ambitious effort being the erection of a wooden hut in the garden of Aberdeen House big enough for grown-up people to enter. When the roof collapsed under undue strains he demolished the hut and used the material to build an effective switchback railway from the drawing-room window to the garden. He explored the whole district round Sydenham on his bicycle to the radius of a day's journey, and occasionally made longer trips with a school friend, spending some days away.

The story of those days would not be complete without a paragraph of secret history, the revelation of which is no longer an indiscretion. Mike was addicted to playing truant from school, and we may assume that he was versed in the art of plausible excuses both at school and at home. He was the leader of a sworn band, other members of which were Arthur Griffiths (" Griff " for short), Ned Sleep and Chris Kay. With such names they could not help playing at the hunt for hidden treasure on desolate islands, the chosen haunt being a strip of private wood adjoining the railway. Many a long day they spent there, cowering in a hollow under the root of a great tree, speaking in whispers, for might not the next lair hide the lurking shapes of Ben Gunn, Black Dog, old Pew, and even Long John Silver himself ?—in that wood in those days time and space, fact and fiction were a continuum of romance. All things there were held in common by the four, and the properties in the drama that was being lived included a revolver with cartridges, an air-gun, a flute, a concertina, and the hull of a large model boat, the rigging and altering of which gave rise to lengthy discussions and very unsatisfactory results. Food was stored up also, for missing school meant doing without dinner, and there was a box of the cheapest cigarettes on the market, which Mike smoked with the best of them, and once when cash was available a bottle of cooking sherry was smuggled in for a grand carouse. This Mike would not touch, and the others before long regretted their rashness. All the talk was of adventure, and many a rousing tale of the sea did Mike read aloud to his comrades, all of whom resolved to be sailors ; and remarkable as it may appear, all four grew up to follow the sea.

Three of the band once, when in funds, bought third returns to London Bridge, and after many rebuffs from policemen at the gates, managed to get into one of the docks and roamed about all day among the ships, trying now and again on a sailing craft to volunteer as cabin boys. This failing, they boarded a big steamer, where the chief steward was engaging his staff of men and boys : here they fell into the queue of applicants ; but when they reached the great man they were greeted with a fatherly smile, and told to hurry home before they were missed. One day in the docks was not enough, however, so the boys walked the streets till late, then slept in a van under the arches of London Bridge Station, and spent the next day by the riverside gazing at the ships going out to all the wonders of the world.

There had been talks at home as to Ernest's future career. His father hoped that he would adopt his own profession and prepare for the study of medicine ; but the boy recoiled from the prospect of years of heavy study followed by a lifetime of dull routine. He was all for freedom and adventure ; the world was so full of possibilities that there must be some short-cut to wealth and fame, some chance at least of a merry life of constant change. The Navy naturally suggested itself, and failing that there was always the mercantile marine. No other career was really open to a boy of spirit. If in the seething mind of the boy, in those years of troubled thought in which school drudgery held so small a place, there had been a recorder as skilled and careful as the form-masters who were shut entirely out from it, we could speak with certainty where we can now only conjecture. But we cannot be wrong in finding great vague thoughts of the future he longed for. Every evening as he followed his lengthening shadow down the long hill from the Crystal Palace on his way from Dulwich College, he might see the image of his mind projecting himself farther and farther beyond the pleasant suburb out into the great world. Had his father been less sympathetic and far-seeing, the boy might very well have run away to sea as the legend ran even in his lifetime. The father of a schoolfellow, a retired master mariner, gave his advice as to how best to begin a sea-going life, for the hope of joining the Navy had to be abandoned, and Ernest approached his father's cousin, the Rev. G. W. Woosnam, later Archdeacon of Macclesfield, who was superintendent of the Mersey Mission to Seamen, for an introduction to a shipping firm. On ascertaining that Dr. Shackleton was willing to apprentice his son if his fitness for a sea life were proved by a trial voyage, and if a good ship and captain could be found, Mr. Woosnam arranged with the North-Western Shipping Company, then under the same management as the White Star Line and flying the same house-flag, to allow young Shackleton to make a voyage on probation, with the rating of a Boy, but with the treatment, uniform, and accommodation of an Apprentice. This left him free to continue to follow the sea, or to give it up at the end of the voyage.

At the beginning of 1890, three months' formal notice of withdrawal was given to Dulwich College, and the last term began for a boy with a new purpose in his life The effect was magical. But for the "disastrous" German, his reports were excellent. In Mathematics he was third in twenty-five—for mathematics are the basis of navigation. This earned from the form-master: "He has given much satisfaction in every way. There has been a marked improvement both in his work and in his behaviour." In History and Literature he was second in a class of eighteen, in Chemistry, third in twenty-seven. At the end of this final school report the Headmaster wrote: "I hope that he will do well." Nineteen years later Ernest Shackleton presided at the prize-giving at Dulwich College, and after receiving many compliments for having done well as an explorer, delighted the majority of the boys by exclaiming that he had never been so near a Dulwich College prize before And on his death the memorial notice in the College magazine opened with the words: "Sir Ernest Shackleton was without doubt our most famous Old Boy." We fear that the school education of the most famous Old Boy had less to do with his success in life than had the influence of his inherited qualities and the literary atmosphere of his home, which stimulated his insatiable love of miscellaneous reading Nevertheless, he retained to his death a deep affection for his old school and its masters.

To some of his reading during his schooldays we get a clue in the books presented to him by various friends. In one of these, Hugh Miller's *Schools and Schoolmasters*, he had marked a passage relating to caves, possibly with the old thought of hidden treasure: "For one short seven days—to borrow Carlyle's phraseology—they were our own and no other man's" Another book read during his Dulwich College days was Samuel Smiles' *Life of a Scotch Naturalist*; the famous Thomas Edwards, whose love of nature might just as easily have made him a poacher as the man of science he was. As a boy Ernest was deeply religious, and keen in the propaganda of the Band of Hope, lecturing on Temperance to the servants, and inducing them to sign the pledge by his irresistible winningness of appeal.

School was over and the sea career was accepted. The old

[*Photo.*] [*Brown, Barnes & Bell.*]
ERNEST HENRY SHACKLETON, AGED 16, IN UNIFORM OF
WHITE STAR LINE.

Face p. 29.

brown chest, which had started down Channel with his grandfather Gavan nearly fifty years before, was hunted out and packed with the modern outfit for the sea. Before sailing he was photographed in the White Star uniform, and the picture here reproduced, with its air of self-confidence and determination, shows that he was resolved to rise to the height of his opportunities; and since his jacket cuffs were held firmly behind his back, who could tell from his countenance that the gold loop and broad stripes of an Admiral of the Fleet were not already sprouting there?

CHAPTER III

THE *HOGHTON TOWER*. 1890-94

"He rose at dawn and, fired with hope,
  Shot o'er the seething harbour-bar,
And reach'd the ship and caught the rope,
  And whistled to the morning star.
. . . .
God help me! save I take my part
  Of danger on the roaring sea,
A devil rises in my heart,
  Far worse than any death to me."
                                TENNYSON

ON 19th April 1890 young Shackleton travelled alone to Liverpool, where he was met by Captain J. B. Hopkins, a friend of Mr Woosnam's, who took him to the North-Western Shipping Company's office and saw him through the necessary preliminaries. Ten days later Mr. Woosnam left the boy cheery and happy on board the full-rigged ship *Hoghton Tower*, a fine clipper of 1600 tons, under the command of Captain Partridge, who was well known in all ports on the great sailing routes as a man of high character and kindly nature. On 30th April the tall ship was towed down the Mersey out into the Irish Sea, and set her sails for the long voyage to Valparaiso.

The new Boy, who had never spent even a week from home except in the houses of relatives, naturally felt the contrast in his new life. The drunkenness of the sailors as they came on board horrified him, their language was no less shocking to one who came from a religious home. Added to this, the strange food, the rough surroundings, and the first lift of the sea might well have quenched the anticipatory enthusiasm for "a life on the ocean wave." And there must have been something like despair in the boy's mind when he gazed at the masts and yards, the great wire stays, and the infinite clusters of ropes coming from unknown attachments aloft and hung in hanks round the base of each mast, every one with a name of its own and a special use, but all unknown and unguessable, like the words in a Russian newspaper. There is no royal road to such learning, and young Shackleton had to be licked into shape on board like any cub of a land-lubber. He knew that he had got his own way and complaint was useless, so he strove to adjust himself to his environment, repugnant as the drudgery of deck-scrubbing and brass polishing might be, doggedly determined to see it through. His quick mind picked up the scraps of information flung at him ; he soon formed friendships and got the help that is never far from the anxious learner, and we may be sure that he was the merriest on board by the time the ship struck the trade-winds and entered the tropics. He was fortunate in his shipmates. The captain was kind and considerate, having the apprentices to dinner with him occasionally, and bringing them into the cabin every Sunday evening for hymn-singing to the accompaniment of his flute, on which the boys thought him but an indifferent performer. The second mate, in whose watch Shackleton was, proved the falsity of first impressions, for though the boy disliked him greatly, in a week or two he recognized him as a real friend, and liked him better than any other officer on board. In a letter addressed collectively to his " dearest Father, Mother, Grandmother, Brother and Sisters," Ernest says that he was afraid he would be laughed at when he said his prayers, " but the first night I took out my Bible to read they all stopped talking and laughing, and now every one of them reads theirs excepting a Roman Catholic, and he reads his prayer-book."

He got over the slight sea-sickness, which was all that fell to his lot, in three days, and on the third day out went aloft to the upper topsail yard, and in a month he was " as much at home aloft as on deck." For many weeks the weather was perfect, the ship raced through the Trades, on one day making 300 miles —a good record even for the steamers which ply in the South Atlantic. The Canary Islands were passed on 13th May ; a huge dead whale was seen, and smelt, a few days later ; and the first flying fish, the first shark, and the torrential rain of the Doldrums were all duly chronicled. On 30th May St. Paul's Rocks were sighted, the first purely oceanic islets that Ernest Shackleton ever saw, and the last, for here the *Quest* was destined to bring him thirty-one years later. That night Father Neptune came on board with the full ceremonial of his court, and the new apprentices were tarred, shaved and ducked on crossing the Line, thus being formally admitted into the freedom of the Seas. In the course of the voyage he was initiated into the noble art of self-defence, and he never ceased to love boxing

The pleasant weather, which had made the hard life of a beginner comparatively easy, broke in June, and for six weeks the *Hoghton Tower* met a succession of furious gales which buffeted her off Cape Horn, blew away many of her sails, smashed some of the lighter spars, carried away two boats, and inflicted serious injuries on many of the crew, the kindly second mate having his thigh broken Shackleton narrowly escaped being struck by falling tackle; but he came through unhurt and was hardened by the experience. Twenty years later he said, in an interview published in *The Captain*. "During my first voyage I felt strongly drawn towards the mysterious South. During that voyage, which constituted one of the stiffest apprenticeships surely that ever a boy went through, we rounded Cape Horn in the depth of winter. It was one continuous blizzard all the way, one wild whirl of stinging sleet and snow, and we were in constant peril of colliding with icebergs or even of foundering in the huge seas. Yet many a time, even in the midst of all this discomfort, my thoughts would go out to the southward, across that great expanse of southern sea, the loneliest tract of ocean in the wide world, the region which seemed to have been especially guarded against the approach of man by the Great Ice Barrier."

It was a blissful thing to reach Valparaiso about the middle of August and catch a glimpse of civilized life again. He made the acquaintance of a pleasant Scottish family with several daughters, " nice sensible girls, who don't fish for compliments though we give them many " The captain took him to dinner at the Consul's, where, he assured his father, he did not take any wine, as he was a teetotaller, but he skilfully completed the sentence by divulging the fact that he had been obliged to take to smoking in the bad weather off the Horn, where such indulgence became a real necessity. The stay at Valparaiso was pleasant ; but on 1st October the *Hoghton Tower* got under weigh and proceeded to the dreary tropical roadstead of Iquique, where she lay, discharging a cargo of hay and taking on a cargo of nitrates by boat, for six long weeks. It was a dismal hole, where it was unsafe to go on shore in the evening, " as the people would think nothing of sticking a knife into you." Moreover, there were no girls who could speak English On the day before the ship left on her return voyage, Shackleton fell into the sea when sliding down a loosely tied rope into a boat, but he was observed and promptly rescued none the worse for the ducking. The experience of handling boats in a heavy surf at Iquique was invaluable in later life, and it could never have been gained by easier methods.

The *Hoghton Tower* got away from Iquique on 1st December and called at Falmouth one day in March 1891, short of food and water, not to end her voyage, but merely to receive orders to proceed to Hamburg. Twenty years later Sir Ernest Shackleton was opening a Flower Show at Falmouth, and recalled his enjoyment on that earlier visit of a full meal of fresh eggs.

It was near the end of April before the hardy young sailor, "grown and changed so," as he put it in a preparatory letter, got back to Liverpool. He had no complaint to make as to the ship, the captain or any one on board, but he had found the disciplined life irksome, and on landing was quite clear in his own mind that he would not go a voyage in that ship again. The captain said to Archdeacon Woosnam in the course of conversation soon after his return, " I expect he has not given you a very bright account of his life, but he is the most pig-headed, obstinate boy I have ever come across, yet there is no real fault to find with him, and he can do his work right well, and though he may not want to come with me again, I am quite ready to take him "

His home-coming was a surprise to his eight sisters, who

34 THE LIFE OF SIR ERNEST SHACKLETON

rushed upon him as he entered the house. For a little while he was so overcome that he could not speak, but he soon recovered and played tricks on the family, pretending that he had forgotten how to eat in a civilized fashion, taking up his chop in his fingers and staring at the knives and forks as at strange implements.

If his friends cherished the hope that young Shackleton had had enough of the sea, they were soon undeceived, and after two months at home he was formally indentured by his father as an apprentice to the North-Western Shipping Company, and made ready for a new voyage in the old ship He had never before enjoyed with such zest the comforts of home, the companionship of his worshipping sisters, or the glimpses of old schoolfellows at Dulwich; but his experiences of nearly a year abroad were limited to the ship, the phenomena of sea and sky, and brief trips ashore at Valparaiso and Iquique.

The *Hoghton Tower* sailed from Cardiff on 25th June 1891, with a cargo of patent fuel for Iquique, under the command of Captain Robert Robinson, a firmer if less genial master than the kindly flute-player of the last voyage. There were seven other apprentices in the half-deck, but only two of them had longer experience of the sea than Shackleton, and he was the senior of the three in the second mate's watch, and thus escaped the worst drudgery. Some minor hardships below he mitigated by forethought, and he set much store by a little brown teapot and spirit stove, which enabled him to make tea, for that supplied on board without sugar or milk was quite undrinkable. A fine cake presented by the Sydenham cook was devoured by his companions while several days of sea-sickness confined him to a diet of milk biscuits.

A long journal letter detailed the events of the outward voyage. There was more work, less fun, and stricter discipline than last time. No Sunday hymn-singing nor dinners with the captain, no visit from Father Neptune even. Shackleton needed all his determination to keep up his Bible-reading, for this time the other fellows, unrestrained from above, were scoffers. He was disheartened by the readiness with which the sailors took the pledge afloat and broke it on shore ;' he could find no one but a negro sailor with whom he could talk of religion except in controversy. Nor was he better off in his literary aspirations . no one else cared for poetry, and the only poetical work on board was a copy of Longfellow. He read it until he knew every line, but he sighed for Milton. On the balance, Shackleton was certainly struggling with far more successes than failures against a hard and degrading environment. Amongst the books he had taken with him was one with readings for every day, *Daily Help for Daily Need*, inscribed " Ernest Shackleton, with love and all good wishes from L. D. Sale-Barker." Mrs Sale-Barker, well known at that time as a successful writer of books for young people, had long been a close friend of the Shackleton family, and her death before his return from this voyage was a real sorrow to Ernest. Another book, Thayer's *Tact, Push and Principle*, bears the inscription : " Ernest Henry Shackleton, with warmest good wishes and earnest prayer for his temporal and eternal welfare from his clergyman and friend Henry Stevens." Such were the influences which reinforced the effects of the home surroundings. He also had with him several of Scott's novels and Thackeray's *Vanity Fair*, which he lent to the first mate, who wanted to read it particularly, as he had heard it was the best written book in the English language. He also read a great deal of history in his early voyages. Motley's *Rise of the Dutch Republic* he never forgot.

His keen observation fastened on every new feature of sea or sky ; he describes his first sight of a fog-bow, and he revelled in the tropical sunsets :

" Many a painter would have given half of what he possessed to have been able to catch the fading tints of the red and golden sunset we had last night. The red and golden gleams gradually fading away into a deep purple, and far away almost on the edge of the horizon was the white speck of a homeward ship. . . . All I say is, if you wish to see Nature robed in her mantle of might, look at a storm at sea ; if you want to see her robed in her mantle of glory, look at a sunset at sea."

The men, recognizing his improved status in his watch, flattered him by calling him the fourth mate, and the second mate roused him to greater interest in the working of the sails by holding out a prospect of being picked as third mate for the next voyage. So two months sped happily on the whole until the ship reached the Horn, and then for a dreadful month observation, literature and ambition were submerged under a desperate struggle with storms and contrary winds on deck, and with invading waters and clashing gear below, making work almost unendurable and meals all but impossible The ship was laid-to for days at a time ; one officer and eight men were disabled by accidents and kept to their bunks, one man was washed overboard in a storm in which no boat could be launched, and Shackleton, too, had to lay up with a fierce attack of lumbago after weeks in wet clothes and a wet bed It is a sailor's superstition that sea-birds are the ghosts of dead mariners watching the ships to see how their successors fare, and we may be sure that if the shades of the three generations of Ballitore were following the *Hoghton Tower* amongst the birds, they would have blessed the courage and steadfastness of their descendant, while it may be some " haughty old albatross cruisin' around," incarnating the spirit of a "turbulent Fitzmaurice" would have had cause to hail some piece of boyish " divilry " or reckless swagger as worthy of a chip of his old block.

It was past the middle of October when the *Hoghton Tower* dropped her anchor off the arid shore and bare mountains of Iquique. Here the old toil of landing and loading cargo by boats went on for two months, made the longer and harder for Shackleton, for he was seized with a bad attack of dysentery, and the coarse talk and disgusting songs of the apprentices who visited him from other ships made him terribly homesick. He welcomed the newspapers which were awaiting him, for politics gave them something decent to argue about.

Of the return voyage of four months there are few particulars available, save for a reference to "terrible weather south of the Horn nearly among the ice," and it was 15th May 1892 when the *Hoghton Tower* re-entered the Mersey. Next day Shackleton reached Sydenham thoroughly tired of the ship, and eager to make the most of his brief holiday in the home where he was so heartily welcomed. Too effusively welcomed indeed, for he found the house beflagged for his home-coming, and passers-by wondering what public event was being commemorated. This he insisted must never be done again on any subsequent return from the sea.

The month passed more quickly than any other of the year. There were old friends to see, old haunts to revisit, cricket to watch at Dulwich College, all the sights of London, with their attractions enhanced a hundredfold, the delights of fresh food daintily served, of the prize roses in the garden, of the old familiar books. His two years at sea had made the youth a stranger on the land. When he accompanied his friends along the streets they had to make him walk next the curb, as without a sharp line to steer by, his gait, habituated to the heaving deck, rolled so that he became a danger to those he met. It was not a month of unmixed pleasure. The ship had grown hateful to him, and he would gladly have changed to another, but the policy of not offending the powers at Liverpool, and the fears of Dr. Shackleton that he might be transferred to a vessel bound for unhealthy ports, induced him to go back to the *Hoghton Tower* and her strict skipper, but he went very reluctantly.

The new voyage was to India, and his friends provided many introductions to people in Calcutta, which were of no avail, for the voyage turned out to be to Madras for orders. His companions in the half-deck were five old shipmates and two new apprentices. The third mate of the previous voyage returned in the same capacity, so Shackleton had no promotion and no more congenial company than before. One of the sailors had been deck-boy on the Irish packet *Banshee* on which Shackleton had made his first voyage to England. The officers were much stricter than on the last voyage from the first day on board, and, experienced sailor as he was by now, Shackleton confessed to being more homesick than ever before. He besought his parents to ask every one to write to him at every port. He sailed on 27th June, the ship having a cargo of salt. He crossed the equator in five weeks, sighting the coast of Brazil a week later; then, encountering a heavy gale in the South-East Trades, the ship rushed through the water at 14 knots for some exciting

38   THE LIFE OF SIR ERNEST SHACKLETON

hours   On 2nd September, when in the neighbourhood of the Cape of Good Hope, a terrific storm struck the ship before sail could be shortened while Shackleton was at the helm. The sails split, the ship, despite all his strength on the wheel, flew into the wind and shipped enormous seas, deluging the cabins and the crew's quarters. She lay absolutely helpless under bare poles ; it was impossible to set the smallest sail or do anything to help the situation · no one expected to see the morning   " All that long dreary night," said Shackleton, "I never heard an oath or any blasphemy from the men. God had frightened them ; they were too near death to swear "

They were half-way across the South Indian Ocean before the bad weather ceased, then turning northward in one fortunate week they logged 1600 miles in six days ; the old ship might have raced a steamer for the time. Madras was reached near the end of October, and the orders were for Chittagong, east of the Ganges-Bramaputra delta, not for Calcutta. The ship lay for two months in the river opposite a little village miles below the town of Chittagong. There was some interest in watching the picturesque costumes of the brightly-clothed villagers by day and the fireflies, " like sparks from a blacksmith's shop," at night. Only one excursion ashore is reported, a rhinoceros hunt with the Commissioner and two other English gentlemen ; but the description of the jungle is not very elaborate, and no mention is made of the bag secured   Most of the time was spent on board working cargo, and Shackleton confesses in his letters that he was sick of the place and sick of the ship. He says that he had written fifty letters since arriving and had only received thirty-six. By every mail he besought those at home to stir up all friends and make sure that he would receive forty letters on arriving at Mauritius—all separate letters, he insisted, and no post-cards. Possibly he felt that his spirits would be raised, not only by the evidence of his friends' continued good-will, but by his comrades seeing that he was so important in the world beyond the ship that he received the heaviest mail of any soul on board. Be this as it may, he went through a period of acute depression at Christmas, and there was no

Photo.] [Temple West.

THE FULL-RIGGED SHIP *HOGHTON TOWER* AT NEWCASTLE, N.S.W., 1893.

Face p. 39.

prospect of relief in the New Year. The future looked to him as dark as the present, and he little suspected that while he was giving way to his nearest approach to despair at the head of the Bay of Bengal, four Dundee whalers had entered the ice of the Weddell Sea in the effort to establish a southern whale fishery. They were doomed to disappointment too; but, nevertheless, their adventure started the shuttle of Fate which was, nine years later, to twine the thread of Shackleton's life with the Antarctic Circle.

Early in January 1893 the *Hoghton Tower* got her anchor up, and was being towed down the river, when an explosion on the tug cast her loose, and the anchor was let go just in time to prevent her from running ashore. Later the strong tide made her drag, and she took the ground, but without damage, and another tug soon got her off and out to sea. In three days she crossed the Bay of Bengal and entered the landlocked harbour of False Point, reputed to be the best in India between Calcutta and Bombay. Here she was to load rice for Mauritius —a horrible job, for every one of the 2600 bags, each weighing 170 lb., that were to be loaded each day had to be passed along the deck from hand to hand, and Shackleton excuses the brevity and illegibility of his letters by the terrible state of his hands frayed by the rough gunny bags. That such a system of loading cargo could exist even on sailing ships in the last decade of the Victorian age, so renowned for its mechanical achievements, is not a little surprising.

The loading over, the ship proceeded on 24th January to Port Louis, Mauritius, where Shackleton had little enjoyment, for he, like many of the crew, fell ill with Mauritius fever, and in spite of consulting a local homœopath, he had to go to sea at the end of March still ill; and his case proved one of the worst on board, he being one of the three still on the sick list when the *Hoghton Tower*, after rounding the south of Australia and passing through Bass Strait, reached Newcastle, N.S.W., on 29th April. Here he speedily recovered, and got into touch with some of his Australian relatives, so that things began to wear a more home-like aspect. The people of Newcastle with whom he came in contact were warm-hearted and hospitable to a surprising degree, and never did they show kindness to one who felt more grateful

Newcastle, N S.W., was far from home, but on leaving it, after a six weeks' stay, the *Hoghton Tower* was still outward-bound, for she received orders to proceed to Chilean ports, and her way lay across the broadest and loneliest stretch of the Pacific Ocean Half-way across, a furious storm descended upon the ship and partly dismasted her. A man had died suddenly the day before, and his body was laid out in the saloon and watched by the apprentices in turn during the night, their vigil being no ceremonial homage to the dead, but the gruesome necessity of keeping the rats away from the body. The superstitious sailors looked on the death as an omen of evil, and they had a gloomy satisfaction in the result. We give Shackleton's own description of the storm

" It was my look-out, which is kept right in the fore part of the ship I had only been there for about ten minutes, and as it began to rain I put on my oilskins and the second mate began to take in the light sails such as royals. It began to thunder fearfully, and the sky was lit up with the most vivid forked lightning when, all of a sudden, a whirlwind struck the fore part of the ship, there was a blinding flash of lightning and a peal of thunder that would wake the dead ; my cap blew off, and I was down to leeward looking for it when I heard a tearing, crashing sound above my head I had just time to crawl up to windward, for the ship was heeling over, when crash down came the fore-royal mast, top-gallant mast, and top-mast ; the heavy wire stays which held them up struck the deck just where I had been a moment before and rebounded into the air about thirty feet It was a miracle that I was not killed. By the next flash I could see that the main-royal mast and yard had also gone, then the rain came down thick and fast and the wind howled and shrieked ; the lightning flashed and played about the rigging. Nature seemed to be pouring out the vials of her wrath on the poor wreck-strewn ship, alongside was top-gallant mast and yards battering about, threatening every moment to knock a hole in the ship's side. At last daylight came and the ship looked a pitiful spectacle, the sails nearly all torn to shreds.

Not only was damage done to the ship, but in the half-deck, the ship being thrown on her side, the chests took charge and one heavy one dashed into my box of curios, smashing coral pipes and seeds, chatties, everything, and what was not smashed was destroyed by water which filled the house ; but I can say nothing, it was a good job we got off with our lives. Directly after we were dismasted all the men and officers came running forward to see if I was alive. . . . For four days we worked clearing the wreck, snatching a wink of sleep whenever we could at night, and after a bit we started to go ahead again, but much more slowly, for we were minus masts and sails, and since then we have been jogging along, first gales then calms."

Two months after leaving Australia the ship was lying in the fine harbour of Talcahuano, Chile, 250 miles south of Valparaiso and the port of the picturesque old city of Concepcion. After discharging some of the Newcastle coal the ship was ordered to Tocapilla, a little seaport about 100 miles south of Iquique ; and here she lay at anchor through the whole of October and November, working cargo by boats. The town offered few attractions and many dangers. The local police had been worried beyond such endurance as they might normally be expected to display, by the conduct of drunken sailors from the ships lying in the bay, and one Sunday evening as he was returning from service in the little Mission Hall, Shackleton fell in with a party from his ship engaged in a tussle with the local authorities. He naturally intervened to get the men off ; but as he knew no Spanish his arguments were quite possibly of a provocative kind and the police turned on him. He ran, they followed with drawn swords, but Shackleton succeeded in reaching the house of a friendly Scot who knew him as a Good Templar and accompanied him to the police office, where he convinced the authorities that the young man was neither drunk nor riotous, and the incident ended with credit to all parties. The horrible monotony of the life at Tocapilla, for such lively scenes were rare, was almost unendurable, and Shackleton got ill again and was greatly depressed.

But at this very time another link in the chain of his destiny was being welded. Since the day following his birth the seas within the Antarctic Circle had been secluded from human sight. Now as he lay despondent at Tocapilla, far away, 3000 miles to the south of him, two Norwegian sealers, who had followed the Dundee fleet of the previous year to the South Shetlands, had succeeded in crossing the Antarctic Circle—Larsen in the *Jason* in the west of the Weddell Sea, Eversen in the *Hertha* on the other side of Graham Land.

At the beginning of December the *Hoghton Tower* got away from the hated anchorage bound for Queenstown for orders, but her ill-luck followed her and she fought for weeks against contrary winds. A violent storm proved too much for her. The masts, perhaps not as strong as those lost earlier in the year, were carried away again, and on the last day of the year she struggled like a wounded bird into Valparaiso Bay. Shackleton welcomed the New Year of 1894, for the long exposure had brought back his Mauritius fever, and in the hospitable British community of Valparaiso he found good cheer and the kindest nursing Mr C P. Brown took him to his own house and treated him more like a son than a stranger Restored to health by the careful nursing and cheered in his heart by feeling himself amongst understanding friends once more, Shackleton reflected with complacency that he had received over 300 letters that voyage and had written nearly 200, surely the champion correspondent of the ship, if not of the mercantile marine ! He had not been the only sufferer. The stern captain had broken down, and was obliged to hand over the command to another and return to England by steamer.

On the way home round the Horn the albatrosses which followed the *Hoghton Tower* one day were probably flying round the Norwegian sealing fleet the next, and all the way home through the Atlantic the ships were within a few hundred miles of one another, but they threw no shadow of coming events.

Before the end of June the ship reached Queenstown, and as her orders were for Dunkirk, and she could not enter that harbour before the high tides of the next full moon, she remained in the Irish port long enough to allow Shackleton to visit his relatives at Moone, and to replenish his stock of clothes, entirely worn out on his two years' cruise. The old sea-chest had been smashed beyond repair in the great gale in mid-Pacific; but he had been working hard at navigation on the homeward journey, and in asking his father for the necessary advance he proudly declared that next year he would be no further expense. The young man had served his apprenticeship and realized his own powers.

He was home again on 3rd July, thankful to have seen the last of his first ship and her coarse and uncongenial company, overjoyed to be with the parents, brother and sisters whom he loved so dearly. On his return, or soon after, his sister Ethel mentioned that she had made the acquaintance of Miss Emily Dorman, to whom she had formed a strong attachment; as a boy he had followed this girl with admiring eyes, and although he had never met her he had not forgotten.

He passed the Board of Trade examination for second mate on 4th October. The long, rough training was over and the work of his life was about to begin. The boy had become a man, and from the troubled waters of his mind the diverse elements of his character were already beginning to crystallize in forms so brilliant but so different, that few who were capable of recognizing and appreciating one of them could understand how the others were also present. Tenderness and sternness, impulsiveness and perseverance, pride in work and pleasure in show, with many others, were all appearing there side by side, and he was beginning to learn how to exhibit them to the best purpose. The obstinacy which his first skipper took to be pig-headed was only the first clumsy exercise of his splendid endowment of endurance which was the glory of his life. His letters now bore the family crest and its never-forgotten motto, "By Endurance I Conquer." If it made him slow to conform to a new discipline, it also enabled him to retain his faith in God and the decencies of life, where many a lad has lost them. As his experience grew he no longer tried to convert the drunken sailors, and his own religion sank inward to be expressed in conduct and kindness rather than in words; but he had got to know the sailor-man thoroughly, and learned how to find the staunchness to duty and simple goodness of heart under the dirty crust; and if he had to give his orders in the words that sailors understand, his own mind was always at home with the poets. He used to say that his communion with the stars when alone on watch in the clear nights at sea so impressed his mind with the austere purity of Nature, that he could find no attraction in the doubtful haunts of foreign ports; the thought of the stars barred the way.

CHAPTER IV

IN THE SHIRES AND CASTLES. 1894–1901

" See the shaking funnels roar, with the Peter at the fore,
   And the fenders grind and heave,
And the derricks clack and grate as the tackle hooks the crate,
   And the fall-rope whines through the sheave.
      It's ' Gang-plank up and in,' dear lass,
      It's ' Hawsers warp her through!'
      And it's ' All clear aft' on the old trail, our own trail, the out trail,
      We're backing down on the Long Trail—the trail that is always new "
RUDYARD KIPLING

ONE day in the autumn of 1894 Shackleton called on an old schoolfellow, Mr. Owen T. Burne, who was in business in London, and said, " Owen, can you find me a job ? You must know a lot of shipowners, and I want a job at sea." They went together to the office of the Welsh Shire Line, where, after some questions, the manager offered a fourth mate's billet on the *Monmouthshire*, which was gratefully accepted. A few days later Mr. Burne met the managing partner, who greeted him with, " That's a rum fellow, Shackleton ! " and in response to an inquiry as to what he had done, went on, " He went down to see the ship, and said he didn't like the fourth mate's quarters, but would go as third ! " Mr. Burne expressed regret, and asked the shipowner what he had done about it, and to his pleased surprise heard, " Oh, I rather liked the chap and gave it him."

Thus Ernest Henry Shackleton found himself on 15th November 1894 steaming down London River as third mate of the *Monmouthshire*, a steamer of 1900 tons register, bound for the Far East. The ships of this company, though always trading to China and Japan, did not run a regular service at fixed dates ; but tramped the seas wherever their cargo could find a market, so that the voyages were uncertain in length, and might be prolonged like those of a sailing ship. Passengers were carried sometimes, and there was a much more varied life on board than in the *Hoghton Tower*. Shackleton probably valued most the seclusion of a cabin to himself after being cooped up for four years with a crowd of noisy and ill-mannered youths. He could read and write now undisturbed during his watch below. He had a good stock of books with him, including poetry, novels, and works of general interest. A neighbour at home had given him Brassey's *Naval Annual*, his father's coachman presented Burke's *Essay on the Sublime and Beautiful*, inscribed, "To Master Ernest from Johnson." Another gift was a popular book on "Famous Men of Science," wherein he had noted that Galileo's birthday was the same as his own.

The Mediterranean opened before him for the first time; he had his first glimpse of the desert in passing through the Suez Canal, and all the interest of the rocky coasts and flaming stars of the Red Sea; and his imagination glorified the harbours and towns of the Straits, China and Japan, so old and rich and mysterious compared with the stark modernity of the nitrate ports of Chile.

While Shackleton was enjoying the pleasant warmth of Indian seas in January 1895, a Norwegian whaler had forced its way through the ice-pack south of New Zealand. This was the *Antarctic*, commanded by Captain Kristensen, with an Australian volunteer named Borchgrevink on board, and they, for the first time since Ross in 1843, were gazing on the great mountains of Victoria Land, and first of all men they had set foot on the soil of the great Antarctic continent. Shackleton could not know of this, but his thoughts were away in the Frigid Zone, for he was hammering out a set of verses entitled " A Tale of the Sea," in which he pictured a long procession of ships of all ages coming out of the north—on each

> . . Nailed to the rotting flagstaff
> The old white ensign flew,
> Badge of our English freedom
> Over all waters blue.

IN THE SHIRES AND CASTLES. 1894-1901

The verse halted somewhat, but the intention was fine, and one stanza bears quotation:

> Then they told me a wondrous tale,
>   And I strove to write it down
> But my pen refused its duty,
>   And I lost my chance for renown.
> But since that vision left me
>   I have looked on those sailor men
> As worthy the brightest idyll
>   That poet could ever pen.

It was not all dreaming. The captain required the third officer to take daily observations for position along with his superiors, and Dr. Shackleton had sent out a sextant and other instruments to meet the returning *Monmouthshire* in April, but the parcel arrived empty. It is amusing to note that Shackleton's letters to his father on this disaster are curt and clear as befits a man, no longer the flowing explanations and lamentations of a boy.

The ship was now bound to American ports, and she visited New York, Newport News, and Fernandino, Florida, where she loaded up for Hamburg, Antwerp, and London. Shackleton had a great deal to do in the ports, checking the cargo; and he relates with pride that there was not a package lost at New York of 4000 tons of the most miscellaneous cargo that ever entered that port, and he had tallied half of it. This was a notable feat, showing unceasing vigilance and accuracy, as any one who has seen the rate at which a steamer discharges can judge, on reflecting that every individual package had to be identified and noted without delaying the work. May and June were passed in the American ports, and late in July Shackleton came back to Sydenham, pleased with his seven months' labours, and bearing proudly a brood of young alligators as domestic pets. He kept these little monsters in the house and garden until the threat of a domestic strike, for the maids shrieked with terror when the creatures smiled at them, caused their transfer to the Zoo.

He had found the captain "a very decent fellow," and he had kept him supplied with newspapers, while he begged his father to send down some of his finest roses to the ship on her arrival, so that the captain and his wife could see what real roses were. As he wrote that, he added, "My goodness! don't I wish that I were a captain with £300 a year!"

He thoroughly enjoyed his little glimpse of life at home—it was only ten days—and never was happier. While he was seeing old friends at Sydenham, the Sixth International Geographical Congress was meeting at the Imperial Institute. Mr. Borchgrevink read a paper on his landing at Cape Adare in the Antarctic, and under the presidency of Sir Clements Markham all the geographers of the world resolved that the time had come for the renewal of Antarctic research on an extended scale. Even as they decided this momentous resolution, Shackleton was leaving the Thames for China and Japan on his second voyage in the *Monmouthshire*

At Nagasaki he bought some books, among them two which show him still pursuing his literary education, one a Rhyming Dictionary to aid his efforts at verse, the other Lemprière's Classical Dictionary, from which to patch his neglected knowledge of the past New Year's Day 1896 found him approaching the Red Sea, bound by Suez and Gibraltar for New York, and on 19th April he was back again in the Thames, where he said farewell to the ship.

The two months which he now had ashore saw him through his Board of Trade examination for First Mate, and he was appointed to the *Flintshire*, a steamer of 2500 tons register, with promotion to the rank of second mate, which showed satisfactory progress in his profession.

His next voyage occupied seven months, and led him round the world by the now familiar Suez route to China and Japan, and across the Pacific to San Francisco, where he spent a month in October–November, thence home by the South American coast, remembered from the sailing-ship days, with a call at Coronel for coal. Putting in at St. Vincent, Cape Verde Islands, for coal on 12th January 1897, Shackleton ran against the old school friend Rhodes with whom he had fought in honour of St Patrick at Fir Lodge, and the meeting made the arid island bloom with happy memories.

IN THE SHIRES AND CASTLES. 1894–1901

He was at home for his birthday for the first time in eight years. Then he was off again to Saigon in Cochin-China with a pleasant lot of passengers ; but the work was growing harder, and his prospects, if he stayed with the company, less bright ; in fact, he felt the restlessness that was always with him the precursor of a change. All the way home he pondered on what the next thing was to be. He thought that he would change his ship for a better ; but when he reached Sydenham in the height of the rose-season of 1897 he made a discovery which drove the thought of ships from his mind. In the drawing-room of Aberdeen House he was introduced to his sister's friend, Miss Emily Dorman, and, if they did not fall in love at first sight when their eyes met, they became instantaneously such true friends that the world was changed.

In the few weeks ashore the new friendship left no time in which to seek a new ship. He had to sail in the *Flintshire* on 17th July for seven long months ; east to Japan, across the Pacific to Portland, Oregon, and home by Coronel and St. Vincent, with an aggravating detour to Marseilles. But on this voyage there was now a nucleus round which his poetic fancies grouped themselves in the night watches under the stars.

He was back in England in February 1898, and though he read of the departure of a Belgian Antarctic expedition, it meant little to him that the *Belgica* was crossing the Antarctic Circle as he was entering the Thames. He was bound for brighter and warmer regions that are not confined within the net of latitude and longitude. There was only a fortnight, but the fortnight was a dream of the new world. Miss Dorman opened up to him the poetry of Robert Browning—he used to say formerly that he did not like Browning, but now the light had come. They visited the British Museum and National Gallery together, and the world of art was opening before him when the call of duty in the harsh tones of the *Flintshire's* siren brought him back to the sea, the old routine, and the commonplace passengers to whom he had to pay civil attentions.

Stirred by a new incentive, he was working for his examination for a Master's certificate, and this he passed at a Naval

50 THE LIFE OF SIR ERNEST SHACKLETON

Court at Singapore on 28th April, rejoicing that he was now, at twenty-four years of age, legally qualified to take command of any ship in the merchant service.

With midsummer he came back to the roses for a month, not so often at Sydenham now as at Mr. Dorman's country home at Tidebrook, Sussex, where the days sped happily, though still the poets and the painters were the intermediaries between a lover and a friend. When the long evenings summed up the glories of the English summer days for Shackleton, away on the other side of the world the *Belgica* was drifting fast in the floe, through the blackness and misery of the first Antarctic night that mankind had ever faced ; but to him there was as yet no Antarctic.

At the end of July the *Flintshire* sailed with a second officer who had left his heart behind him, but was still doubtful of the future. Let it not be supposed that he moped about the deck in a hazy dream—that was not Shackleton. He was alert, watchful, cheery, a friend to all on board and chief organizer of every form of entertainment possible at sea. His anxious heart found rest, and torture, in the poets. He had Swinburne with him and revelled in the voluptuous roll of the verses, and he had now Browning also, the *Selections*, as well as a Browning Birthday Book, copies of which he and Miss Emily Dorman had exchanged and read in daily. This is how he tried to express his feelings :

"The future so uncertain that I dare hardly shape a hope. I do not want Paracelsus's happiness. I would attain, but the goal is that to which Aprile yearned. What can I call success ? a few years' praise from those around and then—down to the grave with the knowledge that the best thing has been missed unless the world's success brings that to pass, and for me it seems a long way off " . . .

How he put it to his companions on board appears from a tribute paid to his memory in the *United Methodist* of 4th May 1922 by Mr. James Dunsmore, who had been with him in the *Flintshire* as third engineer. It was usual for deck officers to ignore the engineers on board ship, or at least to deal with them formally ; but Shackleton chose his friends regardless of convention. Mr. Dunsmore writes of this voyage :

" 'Well, Shacky,' I remarked one evening, 'and what do you think of this old tub ? You'll be skipper of her one day.' 'You see, old man,' he said, 'as long as I remain with this company I'll never be more than a skipper. But I think I can do something better. In fact, really, I would like to make a name for myself '—he paused for a moment or two—' and for her.' . . In my bunk that night I felt convinced that the ambition of that man's life was to do something worthy—not only for himself, but ' for her.' "

Shackleton had not been at home for Christmas for eight years, and he wrote to his father from Port Said that this time he would not be absent from the family party. But Christmas Day found him in the North Sea, making for Middlesbrough, in a typical winter gale. Next day the *Flintshire* ran ashore off Redcar, and Shackleton's stern and resolute performance of his duty on that occasion was the admiration of his companions on board. Personally he hailed the mishap with pleasure, for it gave him an excuse for asking a couple of day's leave to do honour to his father's birthday on 1st January.

New Year's Day 1899, however, acquired a new significance amongst his anniversaries, and though the course of true love was not to run without some obstruction from doubts and anxieties, the day decided him that it was imperative to seek employment which offered better prospects of ultimate independence. He wrote at once resigning his position in the Welsh Shire Line to the surprise and disappointment of the managers. He parted with regret from Captain W. A. Evans, under whom he had been for three years, and to whom he owed much help and advice for which he recorded his gratitude many years later.

On an introduction from the same school friend who had assisted him before, he secured an appointment with the Union Castle Line, and immediately began work on one of the ships in the East India Dock. He arrived on board one January day without an overcoat and carrying one or two books under his arm, and began at once to talk of Browning to his fellow-officers, creating the impression, as one of them says, "that he was distinctive, and a departure from the usual type of young officer. Later on I found he was several types bound in one volume."

While he was busy in the docks making friends with new comrades and learning the ways of the smart mail-ships, which refined upon those of the tramps he had been accustomed to, he heard echoes from the Antarctic which still awoke no answering chord in his mind. The *Belgica* after her long imprisonment in the floe had got free and reached South America, and the *Southern Cross* had deposited Borchgrevink and his party on Cape Adare to make the first wintering that men had faced on Antarctic land.

At the end of March Shackleton sailed for Cape Town on the *Tantallon Castle*, of 3000 tons register, as fourth officer; but, although nominally two steps down from his position on the *Flintshire*, the change was a real advance. He responded to the smartness and dignity of a liner under the blue ensign with crowds of passengers often distinguished and always interesting, and for companions young men ambitious of the highest prizes in their profession. He started with the reputation of being a "poetical sort of chap," and this stuck to him throughout his two years in the Line, for it was true, though only a splinter of the truth. On one of the several ships on which he served and on one of the several voyages he made in each, he came under the notice of a rather pompous captain, who, desiring to score off the poet, addressed him one day when on watch. "Is the glorious orb of day visible, Mr Shackleton?" Instantaneously came the reply, "No, sir; the effulgence of King Sol is temporarily obscured by the nebulous condition of the intervening atmosphere." "Humph," grunted the disconcerted skipper. "Got him with his own tackle," observed Shackleton to his delighted colleagues on the bridge.

He made three voyages to the Cape and back on the *Tantallon Castle* during the year, with a month at home between each, a blissful arrangement when compared with the long, uncomfortable voyages of the past. The frequent opportunities for smoothing the run of his deep, personal ambition produced the result which he had determined on from the first, and the

IN THE SHIRES AND CASTLES. 1894-1901

evening talks on the poets and the stars loosed "the sweet influences of Pleiades," and gave to Sirius and Aldebaran a significance which has escaped the compilers of the *Nautical Almanac*. To the day of his death the stars were always the highest images to him and the tenderest. But there was a long way to go before his position could justify the realization of his hopes : as in most of the crises of his life discretion had the upper hand, and now hard work to win a place and name was recognized as the only way home to the uncharted kingdom.

On one of his stays ashore he became a Fellow of the Royal Geographical Society : on one of his trips abroad he met Mr. Gerald Lysaght, who recognized him at once as one who aspired to great things, and showed unusual character, power and determination, and they became lifelong friends.

The South African War had broken out, and Shackleton was transferred to the *Tintagel Castle*, 3500 tons register, with the rank of third officer, and the duty of carrying troops from Southampton to the Cape The two voyages he made in the trooper kept him busy from 14th December 1899 to 31st May 1900, every hour being filled with congenial tasks by way of duty and recreation. He organized signalling classes for the military officers, and spent hours each day instructing them in semaphoring. He was always flashing out new ideas for the amusement of the men, whose crowded quarters and constant drill tended to breed discontent. He got up concerts and sports, and, as a climax, he stage-managed an impressive visit from Father Neptune with all the traditional features of his own first crossing of the Line, glorified by the more ample resources of the liner. A brother officer declares that on this voyage Shackleton was the life and soul of the ship, always popular with the right class of man, while with the wrong class he had nothing whatever to do. From another source we learn that what he saw of gambling on this voyage decided him to stop playing cards for money, even the smallest stakes. But his chief delight was in the preparation of his first book in collaboration with Dr. W. M'Lean, the ship's surgeon. It bore the title " O H.M S : A Record of the Voyage of the *Tintagel Castle*, conveying 1200 volunteers from Southampton to Cape Town, March 1900," and it was profusely illustrated by photographs, showing all phases of life on board a troopship. Before he returned to London he had secured more than 2000 subscriptions of 2s. 6d. for the book, so that its financial success was assured in advance. An interesting episode is thus related in one of his home letters :

"I knew Kipling was out here, so I wrote a letter asking him to contribute a poem dealing with the voyage, and we could then put it in front of the book. Well, next morning I was looking at the ship on the other side of the wharf when I saw a man in a shabby-looking grey suit with gold-rimmed specs., and the face was the face of Ruddy. I called the doctor, my fellow-author, and told him to go on board and try and speak to him, for he knew the doctor of the ship. He did so, and the next minute was talking to Kipling, so I went across and he came up and said, "Good morning, I got your letter, Mr. Shackleton," in the most genial way, and then said that he had no time at present ; but when he got home (he has gone on the *Tantallon*), and we sent him the proof, so that he could grasp the idea fully if it pleased him, he said, ' I will do my level best for you '"

At the Cape Shackleton heard that Borchgrevink had visited the Great Ice Barrier, found its position was 30 miles south of that recorded by Ross, and that he had landed upon it, and travelled over its surface to latitude 78° 50' S, the nearest point to the South Pole yet attained. He knew, too, that a National Antarctic Expedition was being planned by a joint committee of the Royal Society and the Royal Geographical Society, and he had made up his mind that such an adventure was the thing for him It presented a shorter cut to fame, and ultimately to wealth, than following the sea, and he began to inquire about a possible post on the expedition.

With all these preoccupations Shackleton was always on the alert for any new interest, and on the way home he wrote a letter to the *Daily Mail*, which he ought to have sent to *Nature*, on a swarm of bees which came on board as stowaways at Cape Town on 8th May. "We constructed a nondescript hive, and there content they remained for the next seven days. On the

20th we drew near to Cape Verde, passing about nine miles off, and going that morning to see our little colony, found it flown. Surely instinct must account for the bees' quiescent state during the days when land was far off and their sudden departure as we drew near the coast." And he went on to wonder how the bees fared in their new surroundings, and how Darwin would have liked to know of such a voluntary transfer of species from one habitat to another.

On returning to England at the end of May 1900, Shackleton had five months' leave, part of which he employed with Dr. M'Lean in seeing *O.H.M.S.* through the press and in distributing the volume to subscribers, for they were their own publishers and booksellers. The book was out of hand by August, and a copy specially bound was presented to Queen Victoria and graciously accepted, to the no small gratification of one, at least, of the authors.

Shackleton was making inquiries about joining the Royal Naval Reserve at this time, and the arrangement was nearly concluded when some hitch seems to have occurred which postponed it for nearly a year. This disappointment was more than counterbalanced by a visit to the Dormans at Eastbourne, where they had a house for the summer.

In October he joined the *Gaika* as third officer, and made another trip to the Cape and back. Before he went and after he returned he kept hammering away at all the doors that seemed to lead to the National Antarctic Expedition, the ship of which was growing rapidly on the stocks at Dundee, and a skeleton staff, under Lieutenant Robert F. Scott, R N., the selected commander, was already at work making preparations and keeping up relations with the German expedition, which was to start at the same time and make simultaneous observations in another part of the Antarctic.

The first day of the new century found Shackleton at home ; but on 5th January 1901 he left Southampton as third officer of the *Carisbrooke Castle*, 7600 tons, the largest ship on which he ever served, and in her he made his last voyage in the mercantile marine, a short two months to the Cape and back. On his return he reaped the reward of his continual asking by being appointed as junior officer on the *Discovery* for the National Antarctic Expedition, and he got leave from the Union Castle Company for this service.

This appointment he looked on as the first rung of the ladder of success. He had been a good average junior officer and would have made a very popular captain of a passenger liner, but it is doubtful whether he would have been equally happy in the intermediate sphere of a first officer responsible for detailed routine. The following appreciation from Captain John Austen Hussey shows how he struck his contemporaries while on the Union Castle boats:

"Shackleton was contented with his own company; at the same time he never stood aloof in any way, but was eager to talk—to argue as sailors do, and he was fond of joining in the usual ship banter. He had a quiet drawl in his ordinary speech; but however slow his words, his eyes were bright and his glances quick, his leisurely delivery was belied by the vivacity of his facial expression. When he was on a subject that absorbed his interest or appealed to his imagination his voice changed to a deep, vibrant tone, his features worked, his eyes shone, and his whole body seemed to have received an increase of vitality. When moved to still more intense emotion, or when forced out of his self-quietude, Shackleton showed that determined self-reliant, fearless and dominant personality which later was to make him a leader men would obey and follow unhesitatingly. He was not then the same man who perhaps ten minutes earlier was spouting lines from Keats or Browning—this was another Shackleton with his broad shoulders hunched—his square jaw set, his eyes cold and piercing; at such a time he might have been likened to a bull at bay.

"But, withal, he was very human, very sensitive He quickly responded to his sympathetic nature, and was slow to pass judgment on his fellows. He was just as quickly bored by commonplaces and by futile chatter. He could be calmly satirical without malice, and appreciated a subtle allusion and neatly phrased criticism."

Photo.] [C. Vandyk.
SUB-LIEUTENANT E. H. SHACKLETON, R.N.R., AGED 27.

Face p. 36.

CHAPTER V

THE *DISCOVERY*. 1901-1903

' We left behind the painted buoy
  That tosses at the harbour-mouth;
And madly danced our hearts with joy
  As fast we fleeted to the South.
        .   .   .   .   .

For one fair Vision ever fled
  Down the waste waters day and night,
And still we follow'd where she led,
  In hope to gain upon her flight."
                                TENNYSON.

TO Shackleton the National Antarctic Expedition was an opportunity and nothing more. He would have tried to join just as eagerly a ship bound to seek buried treasure on the Spanish Main, or to scour the Atlantic in search of the Island of St. Brandan. He had no natural affinity for the polar regions, no genius for scientific research; but an overmastering passion possessed him and raised his whole being on a wave of ambition which carried him to, and far beyond, the single goal he had in view. Before the expedition sailed he knew that his object was accomplished; the understanding between him and Miss Emily Dorman was complete, and her father's approval was secured. But his nature was such that he had to convince himself that he was worthy of the bliss beyond the voyage by excelling in every branch of the work before him, and laying the foundations on which in due time he would build an everlasting name for himself— "and for her." The depth and intensity of his feelings were hidden from the world, except the few kindred souls who could interpret the vibrant tones of his rich, low voice in quoting from his favourite authors; no others guessed the motive power of the terrific activity and high spirits which marked him as busiest and happiest amongst the busy and happy crowd on board the *Discovery*.

The National Antarctic Expedition, organized nominally by a joint committee of two learned societies, was really the creation of Sir Clements Markham, K.C.B., F.R.S., president of the Royal Geographical Society He raised, thanks mainly to the munificence of Mr Llewellyn Longstaff, half the cost of the original expedition, while the Government provided the other half He selected from the Royal Navy Lieutenant Robert F. Scott (who was promoted to Commander before he sailed as leader), and Lieutenant Charles W. R. Royds, R N., while Captain Scott selected Engineer-Lieutenant R. W. Skelton, R.N., as engineer, and Lieutenant Michael Barne, R.N., as second lieutenant. The chosen crew were mainly man-of-war's men picked from an enormous crowd of volunteers, and it cannot be denied that both Sir Clements Markham and Captain Scott would have liked to see the expedition purely naval in composition and discipline. But Sir Clements knew the polar regions, and he knew that experience of the ice and of the handling of a sailing ship was necessary for the safety of the ship, so he appointed Lieutenant Albert B. Armitage, R.N.R., an officer of the P. & O. Line, who had recently returned from four years' leave of absence on exploration in Franz Josef Land, as second in command of the expedition and navigator of the ship ; finally, he added as junior officer of the ship, mainly because of his knowledge of sails, Ernest Henry Shackleton, who received a commission as Sub-Lieutenant R.N.R. additional to H.M.S. *President* for service in the *Discovery* on 1st July 1901. The *Discovery* also had a scientific staff of five—Dr. Reginald Koettlitz, who had been on the Franz Josef Land Expedition with Armitage ; Dr. Edward A. Wilson, a brilliant artist as well as a biologist (both of these were medical men) ; Mr. T. V. Hodgson, marine biologist ; Mr. H. T. Ferrar, of Sidney-Sussex College, Cambridge, geologist ; and Mr. Louis Bernacchi, physicist, who alone of those on board had already had experience of Antarctic conditions, as he had been a member of Borchgrevink's expedition in the *Southern Cross*. Bernacchi did not join the *Discovery* until she reached New Zealand; but Mr. George Murray, F.R.S., accompanied the ship as far as Cape Town as provisional head of the scientific staff, and Dr. H. R. Mill went as far as Madeira as instructor in oceanography and meteorology.

The *Discovery* was legally a merchant vessel flying the blue ensign by special Admiralty warrant, and the burgee of the Royal Harwich Yacht Club in place of a house flag. Officers, scientific staff, and crew signed on under Captain Scott as master; but all voluntarily accepted Royal Naval conditions, and copied the naval routine and nomenclature, the saloon being termed a ward-room, and the men's quarters the lower deck. This touch of the theatre amused the scientific staff and pleased Shackleton, who had always a keen delight in make-believe; but it was taken quite seriously by the naval men, and undoubtedly served to maintain the high standard of discipline which prevailed throughout. Captain Scott, following the usual custom of the Navy in small vessels, refrained from exercising his right of living apart from his officers, and became one of the ward-room mess, taking the position of president at the table in turn with the others week about, and imposing no restriction on the freedom of conversation. He was singularly sympathetic and understanding, always keen to add to his knowledge, and concerned to give to every specialist the fullest possible opportunities for pursuing his studies. On deck he was a firm commander with a stern regard for detail, and strict in the enforcement of duty.

The *Discovery* was launched at Dundee on 21st March 1901, and was brought round to the East India Dock in May, where she was fitted out, and stowed with the carefully-prepared supplies of food, clothing and equipment which Captain Scott had been engaged in preparing for a year, all the officers, as they were appointed, taking a part in the work.

The ship had been planned for scientific work, especially in magnetism, and the expedition was designed to co-operate with the German Antarctic Expedition under Professor Erich von Drygalski in the *Gauss*, which was to explore the region south of Kerguelen Land, while the *Discovery* undertook to investigate the Ross Sea, south of New Zealand Later, two other expeditions took up simultaneous explorations in the Weddell Sea : Dr. Otto Nordenskjöld in the *Antarctic*, on the east coast of Graham Land, also sailing in 1901 ; and Dr. W. S. Bruce in the *Scotia*, which set out in 1902. The work of the *Discovery* was thus part of an ambitious plan for the systematic renewal of Antarctic research. Special attention was to be given to exploration on land, and, as the equipment included a balloon, two officers were specially instructed in aeronautics by the Army Balloon Department. Shackleton was one of these, and entered keenly into the spirit of the sport During the summer he also carried out experiments with detonators for firing charges, to be used if occasion arose in blasting a channel through the ice, and on his weekend visits to Tidebrook the villagers were more than once aroused in the night by mysterious explosions.

Early in August the *Discovery* steamed round to Cowes, and when ready for sea she was inspected by King Edward and Queen Alexandra, who showed an interest in every detail of the ship and her equipment, so spontaneous and sincere that all formality was forgotten, and when the King gave an extempore speech of farewell, every man on board was heartened for his future labours by the generous words of encouragement and hope The Queen noticed the fine carnations which Dr. Shackleton had sent to decorate his son's cabin, and, with a hearty handshake to all the ward-room party, the Royal visitors departed. Only then was it discovered that no one had remembered to offer His Majesty the refreshment that had been laid in for him, after discreet inquiries of the appropriate functionaries as to the brand he most affected !

Preparations for sea were rapidly completed, and on 6th August the ugly black hull of the *Discovery*, with her short masts and long yards, was steered through the crowd of pleasure yachts ; and, after dropping the lady relatives and friends, who had come to say farewell to sons, brothers and fiancés, at the little town of Yarmouth, the ship in the grip, for the moment, of silent emotion started on her mission.

Life on board was busy and happy. There was much to learn as to the working of the ship, for steam was more familiar than sails to all the senior officers, and the *Discovery* was intended to economize coal by trusting largely to the wind. New friendships were soon formed and new duties learned. Several of the officers were initiated into scientific observations, and Shackleton undertook to determine the density and salinity of samples of sea-water throughout the voyage. He found the minute accuracy required rather irksome, and was long in grasping the importance of writing down one reading of an instrument before making the next. But his inexhaustible good humour made correction easy, and his determination to excel served him here as in other tasks.

The ship was well stocked with books, from the scientific quartos of the *Challenger Reports* to the dainty duodecimos of the Temple Classics, which occupied narrow shelves fixed to the roof-beams of the ward-room. Only there was little time for reading. Shackleton was soon charged with the supervision of the stores, and there was frequent occasion to overhaul the contents of the holds, as some of the tinned foods suffered in the tropics.

From Madeira in the middle of August to Cape Town and Simon's Bay in early October the trip was uneventful, save for an exciting landing on the desolate island of South Trinidad, where Shackleton was on shore for eight hours on 13th September. On the long stretch across the Southern Ocean the temptation of just a look at the ice-floes proved too strong even for Captain Scott's resolve to waste no time on the passage. The *Discovery* was diverted southward, and on 16th November, south of Australia, she had her baptism of ice on the edge of the floe in 62° 50′ S. Then like a schoolboy, fearful of being caught on a pond before skating is sanctioned, she turned and hastened on to New Zealand, reaching Lyttelton on the last day of November. Three strenuous weeks followed. The ship had been leaking in a way alarming to those unfamiliar with the peculiarities of wooden hulls, and she had to be docked for examination and overhauling. Also, it was necessary to empty and re-stow the holds, and the burden of this fell on Shackleton, whose long experience in the Shire and Castle lines stood him in good stead; and he gloried in reaping this reward for his old drudgery. Stowing a ship with food and equipment for exploring lands where his "will fall the first of human feet" was living poetry for him. The day's work over, the abounding hospitality of the most English of all the Dominions competed with the long letters to Wetherby Gardens, where the Dormans now resided, and the shorter but ever-dutiful notes to Sydenham for the hours that should have been claimed by sleep.

The *Discovery* had a fine send-off from Lyttleton on 21st December marred by a fatal accident to a sailor, who fell from the masthead while sky-larking aloft. The ship put in at Port Chalmers to land the body, and, after adding 45 tons of coal in bags to an already heavy deck-load, she sailed on 24th December 1901, severing the last link with the outer world, as there was no wireless telegraphy when the century was new.

We do not attempt here to tell the story of the National Antarctic Expedition; that has been done in monumental form by Captain Scott, and from another point of view by Captain Armitage, in published works. It is enough to follow Shackleton through his experiences upon it, depending mainly on his unpublished diary, which he kept in rough notes and wrote up at intervals on the typewriter. In doing so, like Scott and Armitage, he follows the fine tradition of British explorers, and excludes all personal criticism, passing over the little squabbles and jealousies that bubble up between the best companions when thrown into the closest contact without any change of society for many months. Such quarrels were fewer on this expedition than on most, and none was serious; in fact, Shackleton says, that when one threatened to break out in the ward-room, the quiet admonition, "Girls, girls!" drowned it in laughter—he had not forgotten the lesson of the doll at Fir Lodge.

The voyage to the ice was pleasant, and the weather finer than is usually experienced in that vexed belt of ocean. Hence the diary deals less with the outside than with the inner world. A like-minded officer of H M.S. *Ringaroona* at New Zealand had presented Shackleton with three volumes of Swinburne.

He was reading another favourite, Stephen Phillips' *Paolo and Francesca*, to Skelton, who professed to despise poetry, but liked this—as Shackleton read it. And once " Bernacchi and I were discussing poetry to-night and the philosophy of old Omar ; he seemed to think it very good, and my opinion is that the lines are beautiful, the translation wonderful, but the philosophy maudlin and unmanly. A finer man in every way is my old favourite Browning, though it is—or was—fashionable to pretend not to be able to understand him ; why, I do not know."

He was reading all sorts of literature, from Madame de Staël to Owen Seaman, but revolted at the sameness of the stories in the popular monthly magazines. Presumably he was referring to what he excluded from his record when he ejaculates one day, " It is a difficult thing to write a diary and keep the rubbish out," for he overcame that difficulty with conspicuous success.

The first week of the New Year, 1902, saw the *Discovery* safely through the belt of pack-ice which frequently keeps a vessel from reaching the open water of Ross Sea for a month or more. Shackleton had his first lesson in ski-running from Koettlitz on a snow-covered ice-floe, but it was not a very happy one. " I think," he said, " my share of falls was greater than the others."

On 9th January they anchored at Cape Adare, and Shackleton had a fine time on shore, helping the naturalists to collect rocks and lichens, watching the entertaining ways of the penguins and photographing all he saw, for " I have an idea to make up for my lack of facility in explanation " by so doing. The day was not long enough for all he had to see, though the sun was now shining continuously, and as the ship steamed southward the grand panorama of the mountains of Victoria Land opened before them. The great white cones of Mount Sabine and Mount Melbourne shone so gloriously on the horizon that many plates and films were wasted on them before it was realized that they subtended so small an angle as to show only a jagged line in the print. Wilson was constantly sketching all that was too far or too fugitive for the camera.

64 THE LIFE OF SIR ERNEST SHACKLETON

On board Shackleton had now complete charge of the catering, and he was much relieved when Captain Scott gave orders to slaughter seals for food, as his effort to give variety to the meals was often harassing.

On 19th January the smoking summit of Mount Erebus came into view ahead, and next day a landing was made on the granite rocks of a promising harbour, since called Granite Harbour, though packed ice kept the ship far away from it. Shackleton was just coming off watch as the party left the ship, and hungry though he was, he joined it, missing his dinner. But he had his reward. He says:

" From the top of the ridge at whose foot we were clambering along a great deal of fresh water was running down into the sea, and it was here that we made our first important discovery I was with Koettlitz, and seeing some green stuff at the foot of a boulder I called him to have a look at it. He went down on his knees and then jumped up, crying out, ' Moss ! ! Moss ! ' I have found moss ! ! ! ' I said, ' Go on ! I found it.' He took it quite seriously, and said, ' Never mind, it's moss ; I am so glad.' The poor fellow was so overjoyed that there were almost tears in his eyes. This was his Golconda —this little green space in the icy South."

In no earlier expedition had anything approaching a bank of moss been found, only a few straggling shoots barely recognizable as of the family of mosses.

Two days later Cape Crozier was sighted, and the long, grey line of the Great Ice Barrier stretching to the eastward. Shackleton was busy helping Armitage with the magnetic observations when not on watch, so could not land. Then for a week of beautiful weather the *Discovery* steamed slowly eastward, along the face of the great ice-cliffs which Ross had compared to the cliffs of Dover. But the majesty of the scene was by no means equal to the vision conjured up by Ross's description. For days at a time the height of the ice-cliffs ranged between 25 and 90 feet, though sometimes it rose gradually to 240 feet, while the water at their base varied between 300 and 500 fathoms deep, showing that the vast sheet of ice was afloat. As the ship kept close in, often within a quarter of a mile, the structure of the ice could be examined, and it appeared to be made up of layer on layer of compacted snow. The worn sea-face was in some parts fretted into caves, in others the ice showed sharp, vertical fractures from which icebergs had cracked away. When the ship was in longitude 170° W. the following entry was made :

" When I went on watch and had been some time in the crow's-nest (the barrel at the masthead) during the eight to twelve watch in the evening, the ice seemed to be a huge berg, some five miles long, and I sent a message down to the captain. He came up and had a look, and then we went down on to the bridge. After a bit I went up again, and saw that it was evidently an inlet into the Barrier, and that it was no iceberg at all. The captain decided to stand to, and at midnight we sounded and got three hundred and four fathoms, with a bottom of blue mud."

Next day they found themselves at the head of a bay in the icy wall, being then about 240 miles east of Cape Crozier. They turned and rounded the peninsula of ice which enclosed the bay, and pushing eastward for nearly 200 miles to 150° W., always kept the Barrier edge in view, except when fog or snow squalls hid it for a time. On 30th January they found the mountains of an unknown land confronting them, and the water beneath them shoaled to 92 fathoms. Here was the end of the Barrier and an impenetrable pack to bar eastward progress, so Captain Scott named his first discovery King Edward VII. Land, and turned the *Discovery's* bow westward again. The diary says :

" The near bare peak that we measured was 1450 feet high, and the slopes which stretched away east and west were over 1000 ; so that is a very definite discovery, and it does seem curious. It is a unique sort of feeling to look on lands that have never been seen by human eye before."

A bay or inlet in the Barrier near the place where Borchgrevink landed in 164° W. was selected for a landing, and the *Discovery* was brought alongside the Barrier edge as if it were a quay at a place where its surface was 15 feet above the water—

a novel sort of quay, for when moored to it the ship had 1800 feet of water below her keel. The place was named Balloon Bight, for here the balloon was landed and duly inflated by the officers and men trained for the purpose; and Captain Scott, exercising his right as leader, had the courage to make an ascent, though he had no previous experience The wire rope which held the balloon captive was so heavy that it stopped the ascent at 600 feet, and he was hauled down safely. A second ascent was made by Shackleton, who had been fully instructed in England. He took a camera with him and procured several excellent pictures of the undulated surface of the Barrier; but he could spy nothing to the south but unbroken ice. The wind got up and no more ascents could be made, so the gas was let out and the costly experiment was over.

By 6th February the *Discovery* was back under the shadow of Mount Erebus, where Captain Scott had been advised that a passage to the south would probably be found. He soon saw that what Ross had charted as M'Murdo Bay was really a channel, and that Mounts Erebus and Terror rose from an island, now named after Ross The *Discovery* made her way along M'Murdo Sound and found winter quarters at the southwest corner of Ross Island, beyond which the Sound was filled with fast ice. A busy time followed making the ship fast for the winter, landing the observation huts, digging out foundations, and building the great hut that was to accommodate all the party if disaster befel the ship. Work was varied with recreation, and Shackleton, to his surprise, found himself out of condition for playing football on the ice; besides, the sailors played the Association game, and he had played Rugby at Dulwich. So, despite his many tumbles, he gave most of his spare time to trying to keep erect on *ski*. At this he was worse than any other; "must practise the more," was his comment. Part of the icy plain on which they exercised was covered thick with the dull ashes thrown out by Mount Erebus. "The whole place had a weird and uncanny look, and reminded me of the desert in 'Childe Roland to the dark tower came.'" So said the lover of Browning.

The 15th of February brought the birthday of Galileo and his aspiring follower, who commented sadly in the diary that he had not had a birthday at home for nearly twelve years ; but cheered up when he found that all such anniversaries were to be kept as special festivals on board, "a day to remember, for they drank my health at dinner . . . a thing I had not had done before."

A few days later he was sent out by the captain on his first sledge journey, to lead a party of three, his companions being Wilson and Ferrar, with the object of seeing whether a practicable route to the south existed between Black Island and White Island, which were visible to the south of winter quarters. The party set off on 19th February with a sledge bearing the flags of the three explorers, Sir Clements Markham having decreed that each officer should have a personal flag. Thus Shackleton had fluttering from the bow of the sledge the Shackleton arms and their appropriate motto, *Fortitudine Vincimus*. Everything was new, all were inexperienced ; but the methods were modelled on those which Nansen had invented and tested in the Arctic. The clothing was warm woollen with an outer suit of gaberdine made windproof, furs being dispensed with except for the mittens. A light tent was to afford shelter on camping, reindeer-skin sleeping-bags to secure warmth at night, and, most important innovation on the gear of the old sledgers of the Franklin Search, the Nansen cooker, with its aluminium cooking-pot and primus lamp secured the maximum of heating power with the least possible loss of heat and the minimum weight. As well as a sledge, the little party also dragged a light Norwegian skiff called a pram, in case of meeting open water. White Island looked as if it were ten miles off, and they expected to reach it in time to camp there ; but in the afternoon the brilliant weather gave place to a snow-blizzard from the south, and at 11.30 p.m. all three were completely played out and were obliged to camp on the sea-ice. In their inexperience they all got frost-bitten slightly, and their twelve hours' struggle made the rich pemmican uneatable, and they had a pretty miserable time. Next morning dressing was a problem, the frozen boots forming a special difficulty. They reached White Island (not ten, but twenty miles from the ship), in the forenoon, but could not find a place to get the sledge up, so they camped again on the sea-ice. They made their way to the island, roped together, and struggled to the summit, a height of 2730 feet. Cold tea had been taken for refreshment, but though carried in their inner clothing, the liquid had become solid. To the south the sea-ice ran smoothly, to the west they looked on a new range of mountains, the continuation southward of those in Victoria Land, and Shackleton took angles to the prominent features, from which Wilson subsequently drew a map. It was 3 a.m. before they got back to the tent. Next morning they marched for three hours to the south and found themselves no longer on sea-ice but on rough glacier ice, part of the Barrier They returned to the tent, and at 9 p.m.

"The wind started from its old quarter, the south, and soon was blowing very hard. Ferrar went off to sleep, but Billy and I hung on to the poles of the tent, for we knew that if they were to go we would be in a bad way. While we were hanging on we were so tired that we both dropped off also, and when we awoke at 1 a.m the wind had all gone."

On the 22nd they made a great march, not direct to the ship, for big cracks had opened in the sea-ice, forcing them towards the rocks east of Cape Armitage, the most southerly point of Ross Island. They left the pram there, and the spot is called Pram Point to this day. Lightened of their loads, they soon crossed the ridge, and were welcomed on board the *Discovery* from their first ice journey, a little thing enough, but a beginning. Next morning at the Sunday service Shackleton asked for the singing of his favourite hymn, " Fight the good fight with all thy might," and for the time thought maybe of the struggle with the powers of the air in a sense more physical than spiritual.

Warnings of coming winter were now beginning. The stars were seen for the first time on 2nd March, as the sun set early enough to allow darkness to divide the days. Shackleton's work was now mainly the weighing out of sledging rations for the last autumn journeys, following the captain's orders to make them up in bags containing three men's rations for one day; but considering this a bad arrangement, as tending to make people eat too much when not hungry: he thought weekly bags would be preferable. When this expedition was dispatched to leave a message for next year's relief ship at Cape Crozier, Shackleton was the only executive officer left on board fit for duty, as the captain was laid up with a sprained knee. So for a time there was no exercise, and the first Sunday revealed another and altogether unexpected pitfall in the path of vicarious duty:

"Had to read prayers this morning. I am not much of a hand at that sort of thing, and I read the Absolution. Billy told me I ought not to have done that, as I was not ordained. I opened at that page, so I suppose that was why I read it. I did not know anything about the ordaining part of the show."

On 23rd April the sun disappeared below the northern horizon at noon, not to be visible again for four months. The winter routine had long been established. All hands continued to live in the ship in their familiar and comfortable quarters, and the sea being solidly frozen and the ship banked up with snow, communication with the land was easy. The great hut was used for special entertainments, and kept in reserve in case of emergency should anything happen to the ship. The darkness did not produce depression or illness of any kind; for the food supply was ample, and the ship had its electric lights in action, although the windmill on which Captain Scott placed much reliance for saving coal was frequently wrecked by the terrific squalls, and at last totally destroyed. Except during blizzards, when all movement was impossible, exercise on shore or on the ice was kept up, and the scientific staff pursued their special studies: the physicist in the magnetic hut, or at the tide gauge on the ice; the biologist at his holes in the ice, through which he worked a dredge and lowered fish traps. The meteorological work, kept up day and night, was taken in hand by turns by all the ward-room party, and the "night out," which fell to each in rotation, was the heaviest and most dangerous task of all. In some cases, when an observer had lost hold of the guiding rope that was stretched from the ship to the meteorological screen a few hundred yards away, he had been known to grope for hours in the drifting snow before he found his way home Wilson and Shackleton had established a special meteorological station of their own on the top of Crater Hill (960 feet), and to this they repaired almost every day; the stiff and dangerous climb in the midwinter darkness and cold being a bracing experience with an edge of danger. Wilson and Shackleton were inseparable friends. "A walk with Billy" is the commonest phrase in the diary. He had the finest mind and the most attractive personality of the whole ship's company, and was a universal favourite, loved and trusted by all; but with the captain and with Shackleton he interwove his life most intimately, finding in their diverse characters elements the most akin to his own. Shackleton was ready to lend a hand to every one. "Muggins," as Hodgson was called for some reason never revealed, found him an unfailing helpmate in digging his dredging-holes through the ice and hauling his frozen ropes. "My hands seem to stand the cold very well. I am able to handle ropes that have been in the water, pulling up fishing lines, without having to put on my gloves, for at least three or four minutes"

On board every moment was occupied The care of the stores and catering filled many hours every day. There were regular debates in the ward-room, at first weekly, afterwards, as the novelty palled, less frequently One evening, "owing to an argument between Bernacchi and myself as to the respective merits of Browning and Tennyson, we had a competition. Bernacchi reading from Tennyson, and I from Browning, on the subjects of war, love, hatred, humour, and religion; the rest of the ward-room being judges as to which was the best. Browning won by one vote."

The great feature of the winter was the monthly publication of *The South Polar Times*, a magazine edited and typed by Shackleton, and illustrated by Wilson with exquisite drawings and water-colours. As the two volumes which appeared have been reproduced *in facsimile,* the outer world has had an opportunity of seeing the publication which fed the conversation of the *Discovery* company for two winters. The editor and artist surrounded their proceedings with an attractive mystery. An office was installed in one of the holds far away from the highways of the ship, and there the two conspirators worked in silence and alone for hours each day. Contributions were solicited and received from the lower deck, as well as from the ward-room. A beautifully carved letter-box was set up, in which contributions could be secretly posted, and many of these were extraordinarily good. The captain excelled at acrostics, Armitage wrote of his north polar experiences, each officer and member of the scientific staff discussed his special study; Shackleton contributed several poems, and Wilson's art was supported by clever drawings from other hands. Of the contributions from the men the most interesting were from the pen of one of the able seamen, Frank Wild, with whom Shackleton, ever impatient of the restraints of etiquette, formed a friendship which endured to the very last day of his life, and developed into years of brotherly co-operation, as the later chapters will declare.

The South Polar Times was too serious as a literary production to permit of the inclusion of all the humorous and satirical contributions which found their way into its letter-box, so after a time a less intellectual sheet, *The Blizzard*, made its appearance, and performed its function of a safety valve with great popularity and success. It has not been divulged to the public, and we treat it with a like reticence here.

So the winter passed, and much to his surprise Shackleton found that there was no time for the extensive scheme of study he had laid out for himself. He snatched odd hours for such books as Bates's *Naturalist on the Amazons* and *Plutarch's Lives*, but the chances were rare, and on 13th June an event happened which drove Bates and Plutarch back into the shades:

"I was called into the captain's cabin this morning and overjoyed to hear that he had selected me with Wilson to go on the long South journey with him in the springtime. . . . He tells me that I am not to talk about the southern journey yet, as it is private, and we have to be examined very carefully for our health."

But, though silence was imposed and observed, the junior officer could think, and he exulted in the thought that he had been chosen for the crowning feat of the expedition.

On 4th July the light was beginning to come back, and at noon there was a glow in the north, making walking on the ice quite easy On the 7th the captain called Shackleton in to his cabin and gave him charge of the dogs in view of the great southern journey. No one on board had any special experience of dog-driving, and the captain had designed special harness for the teams. Now Shackleton was set the task, in no way alarming to a British sailor, of discovering by his own efforts in a week or two the art that takes a northern Canadian years of apprenticeship to master. The result was better than one could have expected ; but it served to strengthen the fine old British tradition which Sir Clements Markham set such'store by, that the best polar draught animals are the human members of the expedition. And in their hearts the *Discovery* people did not believe in dogs.

Early in August the cold came to its worst, the minimum thermometer going down to 52 degrees below zero at the ship and 62 degrees below at Cape Armitage ; but through it all Shackleton was struggling with an idea that was taking visible shape. With Barne as a helper, and the technical assistance of the ship's carpenter, he was testing a new means of transport on the ice which he called his " rum cart." A blast of criticism met him from his comrades, not much more genial than the blizzard Captain Scott in his *Voyage of the Discovery* says : " Shackleton has invented a new sledge, or rather a vehicle to answer the same purpose, much to the amusement of his messmates, who scoff unmercifully The manufacture of this strange machine has been kept the profoundest secret. . . . It was to burst suddenly on our awestruck world, to carry immediate conviction as it trundled easily over the floe, to revolutionize all ideas of polar travelling, and once and for all to wipe the obsolete sledge from off the surface of the snow. . . . It was the queerest sort of arrangement, consisting of two rum-barrels placed one in front of the other and acting as wheels to a framework on which the load was intended to be placed."

The skua gulls doubtless shared the captain's views, and the patriarchs of their community to-day may perhaps scream that they never saw its like for uncouthness and disappointment until motor sledges appeared upon the scene nine years later.

On 22nd August the diary reads :

"This is a red-letter day, for we saw the sun for the first time after an absence of 123 days. Wilson and I went up to the top of Harbour Hill and saw it from there, and we could see a number of other folk, who preferred the flat to the hillside, out on the floe on the same job as we were, looking out for the sun. The sight was grand, and I felt a real joy at the sight of it. One who has not lost it cannot understand how much it all means to see the good old sun. The clouds around were beautiful, and the smoke pall that hangs over Erebus was well lit up by the rays of light from the sun. We then went over to Crater Hill to see the temperature, and found the captain had struggled to the top and had a good view of all the wonders of nature which were spread out before us. He could not understand how it was that I could climb up those hills in my big boots, when he could hardly get on at all in crampons; but there was no difficulty, for I am accustomed to climbing about in the dark, and so this is not hard."

Soon the early spring sledging began, and the captain took Shackleton and Wilson out on many trips for practice, and in preparation for the great adventure On one occasion they went on the sea-ice northward, along the ice cliffs which terminate the steep snow slopes of Ross Island, to the glacier tongue 9 miles from Hut Point, and beyond towards Cape Royds. Another time at the end of September, the sledging trip was to the south, with stores to be left in a depot at Minna Bluff, a rocky point about 70 miles south of the ship. On this journey the Barrier was ascended close to Cape Armitage, and the party travelled over its surface, which was often rough and sometimes reft into deep crevasses, beyond White Island, the scene of Shackleton's first essay. Growing experience and fair weather enabled the party to cover the 131 nautical miles of the double journey in less than six days, the progress made being sometimes 18 miles, and on one splendid occasion,

25 miles in one day. When they got back to the ship Shackleton was so stiff that he kept aboard for two days, much in contrast with his usual activity.

He soon recovered and for a month the old routine continued, working in the holds, walking with Wilson, helping Hodgson at the ice holes, copying out the last number of *The South Polar Times* for the season, till he hated the sight of his typewriter, reading Huxley's *Lay Sermons*, and Browning, "the first time for months." Then for a few days he worked at his nautical astronomy and practised the use of the theodolite for taking latitudes The sailor's familiar instrument for getting the latitude by observing the meridian altitude of the sun or a star is the sextant; but the use of it depends on either having a clear and sharp sea horizon, or a perfectly level surface of some shining liquid, usually called an artificial horizon. The liquid almost always used is mercury, and as mercury freezes at 40 degrees below zero it is of no use in polar cold. Captain Scott introduced the use of the theodolite, which is set level on a tripod stand by means of screws, and so can be used to measure angular altitudes without reference to the horizon, actual or artificial. It was a valuable innovation, well worth the time necessary to acquire familiarity with a new instrument and the extra weight which had to be carried.

The start for the great southern journey on which so much of the success of the National Antarctic Expedition depended was fixed for 2nd November 1902. Captain Scott has described it in his book from the point of view of a leader anxious to do justice to the men he led, and to render an account of his stewardship to the authorities at home. We have to deal with it merely as an incident in the training of an apprentice explorer without responsibility save for the portion of the daily duties which fell to him, and without a say as to the road to be traversed, or the manner in which the problems of travel were to be solved.

Shackleton hoped to reach the Pole and get back again to a *Discovery* afloat, and with steam up ready to bring him back to share in the triumph of the expedition and receive the offer of a comfortable post at home in which to live happy ever after. Nevertheless he had a clear enough view of the possibility of failure, and even disaster, for the night before the start, "I turned in early after writing various letters home in case of anything happening." It very soon appeared that things were not to work out simply or smoothly.

Captain Scott, Wilson, and Shackleton set out at 10 a.m. on 2nd November with three sledges and all the dogs, nineteen in number, and that day they made 12 miles on the way towards White Island. A party, consisting of Barne with eleven men and three sledges, had gone on three days before, with the object of bringing provisions to a second depot south of that which had been laid out a month previously; but they had met with difficulties and were overtaken on the second day out. Next day a blizzard came on, imprisoning the three men in their little tent. Two long days followed, during which they could hardly put their heads outside. It was a bad beginning, but their time was wiled away by reading Darwin's *Origin of Species*. By the 12th they had passed the farthest south reached in the spring journeys, and soon had the satisfaction of being "the first who ever burst" into the wastes of snow and ice beyond the limits of all earlier journeys. On the 15th the last detachment of the supporting party left to return to the ship, and the three were now alone. Mount Discovery stood up on the western horizon, a valuable landmark; but far out on the surface of the Barrier the three men had a hard and increasing struggle. They continued to march their 15 miles or more per day, but the ice was so rough that they could only pull half the loads at one time, so that after going 5 miles they had to return for the other loads, thus making only 5 miles of advance in the day. They passed 79° S. on the 17th, and continued to flounder along through deep, soft snow, the noontide sun sometimes too hot, the wind always too cold.

As a specimen of the sort of life we may quote three consecutive days:

"*25th November*.—Five miles to the good, 15 miles done. To-day we passed the 80th parallel, so are inside the magic circle, still going S.S.W. I have become a permanent ." hoosh " cook, Billy breakfast cook. We now have no hot lunches, eating our lunch on the way back from the sledges, one piece seal meat, eight lumps sugar, one biscuit. Don't feel up to writing to-night, have a touch of snow blindness, seeing double

"26th November.—No journey to-day; heavy drift the dogs being played out and needing a rest, we did not start. Had only two meals to-day to save food, so read some Darwin for lunch. Weather cleared up in the evening, and we got bearings

"27th November.—Started the dogs off with full load, only got 1 mile with it, then had to take half the sledges on and come back for the others Snow was too much for the dogs. Captain stopped behind, Billy and I went back for other sledge; a cool breeze reducing the temperature to zero pulled the dogs up, and we soon got the sledges up to the camp, where everything was ready for supper as soon as we had changed our footgear. Billy had made some sketches of the distant land in S.W., towards which we are making. We have now decided to steer S W, as we are having such heavy work with the loads, and hope to reach the land sooner than steering S.S.W. Land must be about 60 miles off. We did only 4 miles to the good to-day, covering 12, but there is the satisfaction that every foot of ground is new "

So it went on day after day, " it is drag, drag and drive, drive from the time we get up till it is time to turn in "; " feeling rather tired, hoarse with shouting to the dogs " One day a 4 oz. weight was found amongst the seal meat. " There is a sort of irony about this," when every ounce of burden is limited to strict necessity.

They struggled on, the dogs began to die, the snow grew softer, the men were in a chronic state of hunger; one day the advance was only 2 miles On 14th December they made a depot of everything that could be left behind in latitude 80° 30' S, and here they tried to approach the land, now close at hand, barred by a vast chasm which they could not cross even with the lightest loads, and they were obliged to change their course to due south, parallel with the lofty mountain range with peaks rising to between 7000 and 11,000 feet, on each of which Captain Scott fixed the name of a living admiral. One or another was now always suffering from that intense inflammation of the eyes euphemistically termed "snow blindness," and one day in a fog they blindly crossed a tremendously deep crevasse without seeing it, though the snow bridge which carried them was the only narrow crossing place in its whole length. They struggled on past the 81st parallel of latitude, and they hoped to reach 82° S. before turning—there was no longer a chance to reach the Pole. They were saving food now to have a full meal on Christmas Day.

Hunger was constant. " We always dream of something to eat when asleep. . . . My general dream is that fine, three-cornered tarts are flying past me upstairs, but I never seem able to stop them. Billy dreams that he is cutting huge sandwiches, for somebody else always. The captain—lucky man—thinks he is eating stuff, but the joy only lasts in the dreams, for he is just as hungry when he wakes up " As to the feast:

"*Christmas Day.*—Beautiful day, the warmest we have yet had—clear blue sky. We have made our best march, doing to-day 10 geographical miles, the surface in grand condition. Now we are entirely doing the pulling, the dogs being practically useless. Started breakfast at 8.30. Bill cook.
Christmas breakfast —a pannikin of seal's liver, with bacon mixed with biscuit, each, topped up with a spoonful of blackberry jam; then I set the camera, and we took our photographs with the Union Jack flying and our own sledge flags. I arranged this by connecting a piece of rope line to the lever. Then four hours' march. Had a hot lunch. I was cook—bovril, chocolate and Plasmon, biscuit, two spoonfuls of jam each. Grand!! Then another three hours' march and camped for the night. I was cook, and took thirty-five minutes to cook two pannikins of N. A. O ration and biscuit for the hoosh, boiled plum-pudding, and made cocoa. I must, of course, own up that I boiled the plum-pudding in the water I boiled the cocoa in, for economy's sake; but I think it was fairly quick time The other two chaps did not know about the plum-pudding It only weighed 6 oz., and I had it stowed away in my socks (clean ones) in my sleeping-bag, with a little piece of holly which I got from the ship. It was a glorious surprise to them—that plum-pudding—when I produced it. They immediately got our emergency allowance of brandy so as to set it on fire in proper style; but when the brandy was uncorked it was found to be black from corrosion in some manner, so was useless, and had to be thrown away. We turned in really full to-night.

We have definitely settled our farthest South to be on the 28th, as examination shows that the captain and I have slight scurvy signs. It will not be safe to go farther."

They pushed southward, however, until the last day of the year. Wilson had to go blindfolded on account of "snow blindness," Shackleton was leading for two days. The last camp was about 8 miles from the western mountains, opposite a wide valley leading westward, to which Captain Scott gave the name of Shackleton Inlet. Beyond it rose two great summits, named Mt. Longstaff (over 10,000 feet) and Mt. Markham (over 15,000 feet). The latitude was 82° 15′ S. A gallant attempt was made to reach the great red cliffs to the west, but after descending to the bottom of a big snow gully with no little difficulty, a vertical ice cliff with overhanging summit more than 70 feet high barred them from the land of promise.

Shackleton's diary does not record a tragedy of the last camp, his upsetting of the hoosh-pot, and the awful silence that followed when it seemed that some of the precious food was to run off the tent floor on to the snow Even the captain in his faithful narrative failed to say how the silence was broken. But the food was eaten from the floor.

The return journey had to be hastened, for there was just enough food on the sledges to last the party for a fortnight, and the distance of the first depot required an average of 7 miles a day to reach it in the time This was done. The dogs died or were killed to feed the survivors. The men struggled on, Shackleton going blindfolded for two days when "snow blindness" struck him; but all pulling steadily. When they left the depot with renewed supplies of food on 14th January 1903 the signs of scurvy were strong on all three, and a course was laid straight for the ship.

That very day Shackleton had an attack of hæmorrhage during a fit of coughing, and he was not allowed to pull the sledge any more nor to do any of the heavy camp work. But he continued to walk the 9 or 10 miles of the daily march He acknowledges in the brief entries in the diary that he "was not very well," and he constantly worried about the work done by the others when he was helpless to assist. They tended him with the devotion of true friends. A fortnight brought them to the second depot, and food was then abundant. The old familiar hills about Winter Quarters came in view. Shackleton grew worse daily; but he always struggled on and never collapsed, nor was he carried on the sledge, as rumour later declared. Captain Scott publicly denied the report when it appeared in a London paper; nevertheless Sir Clements Markham posthumously repeated it in his *Lands of Silence* in 1920, to the grief and indignation of Shackleton. The long nightmare ended on 3rd February, when Skelton and Bernacchi came out on the ice to meet the returning party and helped them back to the *Discovery*. Shackleton's last entry runs:

" I turned in at once when I got on board, not being up to the mark, after having a bath—that is the first for ninety-four days. It is very nice to be back again, but it was a good time."

The relief-ship *Morning* had arrived with letters from home. But the pleasure of them was marred by the captain's decision, based on the doctor's reports, that Shackleton should be invalided home, and Lieutenant G. F. A. Mulock, R.N., of the *Morning*, kept in his place. It was the bitterest disappointment of his life. He felt certain that his attack of scurvy was not really more serious than that of the others, and that a month's rest would make him as fit as they would be. But he was learning that the naval discipline he assented to so cheerily admitted of no "reason why," and he went on board the *Morning* with an aspiration, soon to harden into a determination, that he would yet prove to the Fleet and to the world that he was a fit man, perhaps even the fittest man, for polar exploration.

Drawn towards home by tender ties, he yet felt more strongly drawn towards his old companions on the *Discovery*, who were ready for another year of adventure and the hope of great things to be found beyond the western mountains, where Armitage had discovered the existence of a vast plateau 10,000 feet above the sea. On 28th February he sailed on the

Morning, and his lingering gaze on Mount Erebus interpreted the semaphoring of its smoke banner not as Farewell, but as Au Revoir.

On 19th March he landed in New Zealand a sound man once more, and he was not a little consoled by the hearty welcome he received from every one there. For several weeks he worked diligently at Christchurch on matters connected with the stores for the next year's voyage of the *Morning*. After completing this task he passed a few days with new friends who were spending the summer in the wild solitude of the Otira Gorge. His hosts could never induce him to talk of his own adventures in the Antarctic; but they got their information on this subject at second hand from the children, to whom he opened out in the confidence of walks in the bush.

He sailed from Auckland on 9th May in the *Orotava*, bound for San Francisco, on the way to New York and home.

CHAPTER VI

SHORE JOBS. 1903–1906

> "Therefore from job to job I've moved along,
> Pay could not hold me when my time was done,
> For something in my head upset me all,
> Till I had dropped whatever 'twas for good,
> And, out at sea, beheld the dock-lights die,
> And met my mate—the wind that tramps the world"
> RUDYARD KIPLING.

THE annual conversazione of the Royal Geographical Society, held in the Natural History Museum at Cromwell Road, was, during the presidency of Sir Clements Markham, one of the most brilliant gatherings of the London season. Travellers who had faced the exhibits, while yet alive in their native haunts, governors and ex-governors of every state and province of the Empire, and admirals who, having left the sea, could still catch some savour of the great times of the past amongst the wanderers of to-day, were always numerous enough to allow the thousand or so of ordinary Fellows of the Society feel the glamour of the world, for to all of them it was "our conversazione." Amongst such a crowd Shackleton, though not alone, wandered very much a stranger on 17th June 1903, a day or two after his return to England. "That can't be Shackleton," was the usual comment when one pointed out his great athletic figure and bronzed countenance with eyes flashing happiness and fun. "Surely *he* was never invalided home!" And every one who had the opportunity heard, delighted, the first word-of-mouth news of the great Antarctic Expedition.

Amongst those who listened to what he had to say that evening and later, two men were keenest in their interest in the

Discovery and her discoveries. One of these was Sir Joseph Hooker, whose memory of sixty years before, when he was assistant surgeon on the *Erebus*, remained so vivid that he identified point after point in the photographs brought home by Shackleton as familiar landmarks in the great voyage of his youth. The other was Sir John Murray, then president of the Royal Scottish Geographical Society, who had been the junior naturalist of the *Challenger* thirty years before, and had but recently finished the compilation of the fifty huge volumes which comprised the Reports of that expedition. They recognized kinship with the young lion of the Antarctic, whose premature return had made him the first to be greeted with the praise due to the expedition. He made a point of seeing the relatives of all his comrades, conveying the good news of the happiness of the wintering and the strenuous joys of the sledge journeys.

Greetings over, the question of his own future became his main concern He would gladly have joined the Royal Navy, the merits and prospects of which had been well advertised by the ward-room mess of the *Discovery*, and he hoped that the Admiralty might see in his voluntary service under the King's regulations ground for transferring his commission from the Royal Naval Reserve to the Royal Navy itself. The Admiralty were only willing to credit his service on the *Discovery* by promotion from Sub-lieutenant to Lieutenant R.N.R. on his obtaining a test certificate for drill. That would have been useful had he thought of returning to the Union Castle Line ; but he had had enough of the mercantile marine, so did not qualify, and, in the following year, resigned from the Naval Reserve.

For a few months he was kept busy at the congenial work of seeing to the stores and outfitting of a second relief ship for the National Antarctic Expedition. Sir Clements Markham had made a perhaps too impassioned appeal for funds to send out the *Morning* again in case the *Discovery* was not able to break out of the ice in the following February. He got the money, and the *Morning* was dispatched ; but the Admiralty took alarm at the idea of so many naval officers and men being perhaps lost to the service for another year, and they, altogether irrespective of the committee, purchased a fine large sealer, the *Terra Nova*, at Dundee, and employed Shackleton to assist Admiral Aldrich in getting her ready under extreme pressure of time. So Shackleton enjoyed some busy weeks at Dundee, where he was quite at home, as he had been there often when the *Discovery* was building, and before the end of August the *Terra Nova* was off, with orders from the Admiralty to desert the *Discovery* if she could not be got out and to bring the whole expedition home. Never was a polar ship so honoured as the *Terra Nova*, for she was towed by successive men-of-war to Port Said, where she passed through the Suez Canal, and then on to Aden, whence she proceeded under her own steam. At the same time Shackleton also had a hand in the equipment of another expedition, which was being sent out by the Argentine Government in their gunboat *Uruguay*, under the command of Captain Irizar, then naval attaché at the Argentine Legation in London. The object of this expedition was to rescue Dr. Otto Nordenskjöld and his companions of the Swedish Expedition which had wintered for two years in the north-west corner of the Weddell Sea, and whose ship, the *Antarctic*, had been caught in the terrible pack ice of that sea, and after being crushed had gradually settled down and sunk on 12th February with her flag flying.

In September the British Association met at Southport, and Shackleton showed the slides of the *Discovery* expedition to a large and delighted audience in Section E, Geography. The pictures were novel and fascinating, but the charm of the meeting was the artless sincerity of the speaker, then quite unversed in platform tactics, but telling his tale in the words that came to hand. One of the slides, unnoticed by Shackleton, began to melt at the edges, and the operator quietly slipped it out of the lantern to allow it cool; but the lecturer's eye, caught it just disappearing, and, quite forgetful of the audience, he hailed the lanternist in a tremendous voice, as if he were ordering a boat back to the side of a ship—"Hullo, you there! Bring—that—SLIDE—BACK!" Probably he never gave so delightful a lecture again, for he lost sight of everything but his subject.

Talking about the *Discovery* was very fine, but it was necessary to find some remunerative employment, with at least promise enough of a future competence to justify marriage in his sanguine mind. While in the Antarctic he had often spoken of journalism as a career, and expressed some belief in his fitness in that way to make a fortune—a living was too meagre a prospect ever to dominate his plans. Now he secured a position in Sir Arthur Pearson's office as sub-editor of the *Royal Magazine*, and for the first time in his life he set himself seriously to work on shore. During the autumn of 1903 he was attending daily with a scrupulous exactitude as to hours at Henrietta Street, Covent Garden. He took to the life with infinite zest, and in a very short time he endeared himself to the whole staff of the office, for they recognized in him not the sort of man who usually wrote things, but a living specimen of the sort of man they were always writing about—the man who did things. His literary style in those days was far too high flown, and had to be deflated before publication. He was careless in proof reading, nevertheless Mr. P. W. Everett, under whom he served, could write.

"His knowledge of the technical side of bringing out a magazine was *nil*. But five minutes' conversation with the man was enough to show me that he possessed what was far more useful than mere technical knowledge, which any one can acquire, namely: fresh ideas, interest, enthusiasm, and keen journalistic insight.... He was the most charming of men. There was not an ounce of 'side' in him; he was 'hail-fellow-well-met' with every man he came across. I never met a more exhilarating man, a more genial, a better companion, a racier raconteur.... I can see him still, his face, heavy and stern in repose, all alive and lit up with the excitement of retailing his hairbreadth escapes by field and flood. I can hear the deep, husky voice rising and falling with the movement of his story, and sometimes raised, by way of illustrating his point, to a rafter-shaking roar."

He wrote an account of the first year's work of the *Discovery* for *Pearson's Magazine*, and with the reckless generosity that he could never overcome, gave the whole proceeds to the *Discovery* Relief Fund. He was a member of the Royal

Societies Club, and often lunched there at the round table frequented by the officials of the Royal Geographical Society in the days when 1 Savile Row was the hub of the geographical universe. There he met the latest travellers from every quarter of the globe, and talked over schemes for future exploration and the regeneration of the world. At this time he was enthusiastic about a proposed international news agency, promoted by a plausible foreigner, which was to collect the truth, and nothing but the absolutely uncoloured truth, from every country, and transmit it in pellucid purity to the grateful Press of the world. This alluring vision sweetened life for him for a year or more before it faded to leave room for a new excitement.

The path to fortune through journalism could only be traversed slowly, and something more promising took its place. The return of the German expedition in the *Gauss*, after vain endeavours to extend its small discovery of Kaiser Wilhelm Land south of Kerguelen Island, stimulated interest in the Antarctic, and in November Shackleton was asked to lecture in Dundee and Aberdeen for the Royal Scottish Geographical Society. He was well received in Scotland, and he found that there was a vacancy in the Secretaryship of the Society in Edinburgh, which he was quick to see offered better social prospects, if no more pecuniary attractions, than London journalism. He soon made up his mind that this was the thing for him " and for her." Indeed, when he first interviewed the honorary secretary, he frankly declared that he must have a job, because he wanted to get married, and he thought he would like this job.

The formal application was sent in on 4th December, and two days later he met the Recommendations Committee in Edinburgh, was duly selected, and at the first Council Meeting in January 1904 he was formally elected Secretary and Treasurer at a salary which was more nicely adjusted to the resources of the Society than to the dignity of the office. The Royal Scottish Geographical Society was founded in 1884 at a time when Edinburgh was a great world-centre of geographical activity, being the headquarters of the *Challenger*

Commission, and of various institutions primarily religious, scientific or commercial, in which geographical knowledge was of high importance. After a career of twenty years the main body of the Council remained the same, not only as regards the interests represented, but even individually. Were it not for the quick and open mind of Dr J. G. Bartholomew, the great cartographer, who had filled the post of honorary secretary from the first, the natural calmness of maturity might conceivably have barred the Society against the spirit of innovation. But Dr. Bartholomew was supported by a few open-minded councillors, and Sir John Murray was the president of the year, so that the dignity and decorum which had been fostered by the late secretary, who had been placed upon the Council, were felt to be secure. The course of a learned society in Edinburgh had always been grave and ceremonial, black coats alone appeared at the Council table, discussion was deliberate, and resolutions were framed with the formality of a court of law. Hence, when Shackleton lounged into his office on the first day in a light tweed suit, smoking a cigarette, and greeting his assistants with a joke, there was some natural shaking of heads There would have been more if an incident of the early days had come to the knowledge of members.

One morning in February he and Miss Dorman, who had been busy with the decorators and furnishers of their future home, came into the meeting-room of the Society, then the ground floor of the National Portrait Gallery in Queen Street, which was separated by heavy curtains into two rooms for office work, they found an assistant practising with a golf-ball, which he was driving from the far end of the room into the curtain Instead of rebuking his subordinate, Shackleton borrowed the club, and tried his prentice hand at the game. He drove a ball through a pane of the window, right across the street, and far into the Gardens on the other side.

He put the same energy with the same lack of conventional reserve into work as into play On 25th February he met the whole Society and a large part of the Edinburgh public face to face in the great Synod Hall, where he delivered a lecture on "Farthest South," and established in that one hour a reputation as a good speaker and a cheery friend which was never belied. Very soon he was chuckling over having secured twenty-five new members and £125 worth of new advertisements for the Magazine. Almost before the astonished Council realized to what they had given their consent, the telephone was installed, a typewriter was clicking gaily, and a formidable labour-saving engine, termed an addressograph, was doing a week's output of addressed covers in a couple of hours. Within three months of his appointment he had awakened the Society to the full activity of its early years ; and so keen was he to prevent it from settling into its former calm, that he even persuaded his bride to dispense with a honeymoon.

The day to which he had been looking forward through seven years, with a fervour that made them in retrospect seem as one day, was Saturday, 9th April 1904, when at Christ Church, Westminster, Ernest Henry Shackleton and Emily Mary Dorman were married. Mr. Cyril Longhurst, the secretary of the *Discovery* expedition, was best man, and all the friends of Antarctic exploration and explorers then in London were present, all believing, as his parents and sisters believed, after the first voyage on the *Hoghton Tower*, that he had had enough of wandering and was now "to live happy ever after."

The newly-married pair spent Sunday in Peterborough, and on Monday evening they were home and enjoying "a glorious picnic" in their little house, 14 South Learmonth Gardens, Edinburgh, indifferent to the fact that the electric lighting was not in order, and rejoicing that the new servants had not yet arrived. The house is situated at the extreme north-west corner of Edinburgh, not a house beyond it, and it looks straight across the fields to Fettes College and beyond it to the Firth of Forth and the coast of Fife, with the twin peaks of the Lomond Hills and the long ridge of the Ochils, over which in clear weather one can sometimes catch a glimpse of the dim, blue buttresses of the Highlands.

Mrs. Shackleton brought not only the charm of her personality to the help of her husband in his new position, but

88 THE LIFE OF SIR ERNEST SHACKLETON

also introductions from old friends, which opened every door and ensured a welcome from the most exclusive circle of social life in Edinburgh. To many who view this circle only from without, social Edinburgh has seemed but "an east-windy, west-endy place," as Professor John Stuart Blackie put it long ago; for to the outsider the circle is apt to appear as frigid as the Antarctic, meeting the world in a ring of cold shoulders draped with the sombre gowns of judges, church-leaders, professors, and advocates, with here and there the stiff back of an officer from the Castle. Within, it was in the opening years of the twentieth century warm with kindliness, bright with cultured intelligence, glowing with humour and flashing with wit, retaining still much that was best in the Edinburgh which was to Sir Walter Scott "mine own romantic town." Here Shackleton was in a new world which gratified him as he pleased it, and he made new friends who held the keys to many locks. He had opportunities of familiar conversation with statesmen like Lord Rosebery, men of science and culture like Professor Crum Brown, leaders of the business world like Mr. William Beardmore, guardians of the social charm of bygone days like Mrs. W. Y. Sellar, and luminaries of the Bench, the Bar, and the University, who were deep in all the movements of the time. Rarely has so dazzling a transformation befallen the third mate of a steamer in the course of three short years.

The shadow of the Antarctic lay athwart the dial of his life even in this bright and busy year. Dr. W. S. Bruce, the first British subject to promote and carry out an Antarctic expedition on his own responsibility, had spent two summers in the Weddell Sea, into which he had twice pushed to very high latitudes. In the first season he had reached the point where Sir James Ross had reported an oceanic depth of 4000 fathoms and proved that the real depth was 2660. In the second season he got to a higher latitude than was ever before reached in those waters, and found a new coast hidden in the mists which he named Coats Land, after his most munificent supporter; greatest feat of all, as the future proved, he was able, thanks to the splendid ice navigation of Captain Thomas

Robertson, to keep the *Scotia* free and to bring her safe out of the reach of the drifting floes. In July the *Scotia* returned to the Clyde, and Shackleton had the pleasure of carrying out arrangements for the worthy reception of the men his Society delighted to honour. The gold medal of the Royal Scottish Geographical Society was presented to Dr. Bruce by the President, Sir John Murray, when the explorers landed in triumph at Millport.

The summer was spent in Dornoch, trying to learn golf, with an interlude at the British Association meeting in Cambridge. Then came another moving ceremony of welcome. The *Discovery* had lain in her cradle of ice for a year after Shackleton had left her, and in the second summer Captain Scott had made a journey of 300 miles across the great ice-covered plateau beyond the western mountains. The relief ships *Morning* and *Terra Nova* had appeared, and with a breaking heart Captain Scott was carrying out his orders to transfer all collections to them and desert the old ship when, with one last tremendous effort, he had freed the *Discovery* from the ice. He brought her back to New Zealand, and then across the Southern Ocean, through Magellan Strait, to England. On 16th September Shackleton took part in the festivities which greeted his old comrades in London, a lunch on the deck of the *Discovery* in the East India Dock, and a great dinner by the Royal Geographical Society. Later he shared the honours showered on the expedition which had inaugurated Antarctic land travel—the King's award of the Polar Medal, an octagon hung from a white ribbon, and a silver replica of the special gold medal presented by the Royal Geographical Society to Captain Scott.

During the autumn he brought out a Prospectus and Programme of his own Society's lecture session, beautifully got up and illustrated in a way which took away the breath of the older-fashioned members of Council.

In November, the Twentieth Anniversary Banquet of the Society was graced by the presence of Captain Scott, Dr. W. S. Bruce, and Sir Clements Markham, who stayed with the Shackletons for several days. In the course of the month the two Antarctic leaders lectured to the Society on their expeditions at Edinburgh, Glasgow, and Dundee. In December Shackleton himself gave a Christmas lecture to an immense audience of children. So little removed was he from the spirit of youth that in acknowledging an uproarious vote of thanks he said—so we are assured by one of the audience—" Now, you kids, I'll put you up to a good thing. If you want to see what sledging is like, go home and harness the baby to the coal-scuttle and drive round the dining-room table"; and after the laughter, "but don't tell mother I told you!"

So ended a year memorable in the history of the Society, more memorable in the history of its secretary, who was already feeling for the next step. He had had much talk on the politics of the day, and his views, so far as he had formed views, appeared to the Liberal-Unionist politicians to be sufficiently near those by which the party steered to justify overtures being made as to his willingness to stand for Parliament. The possibilities were alluring to an ambitious young man, the party agents suggested his name as a candidate for Dundee, a two-member constituency so stubbornly Radical that there was no real question as to the result except as to whether a Labour member might beat one of the two Gladstonian Liberals who would otherwise be almost certainly returned. To a person not shackled in party politics it would appear just possible that a vigorous, impetuous, and popular Unionist might have a sporting chance of winning one of the seats; but to run two Unionist candidates coupled together was merely a demonstration to encourage the party in other fields. In Dundee itself such opposition could not be expected to produce any other result on the election than to spur the Liberal and Labour organizations to greater activity.

Early in the autumn Shackleton had seen the Liberal-Unionist chiefs in London. They recognized the good fighter he was, and early in January 1905 he appeared before the local party organization in Dundee, was duly approved, and formally adopted as their candidate for the next general election. Shackleton viewed it as a fine adventure and a tremendous lark, and he was quite unprepared for the way in which the news affected the Council of his Society. For one thing, the traditional view of the secretary of a learned society was that he should within reasonable limits be all things to all men and should carefully abstain from wounding the susceptibility of any member. Hence both Unionists and their opponents were at one in deprecating the appearance of a member of the paid staff in the political arena; besides, such a thing was absolutely unheard of, and the unprecedented must be held to be bad until the lapse of much time showed that it might be otherwise. The Liberals in the Society naturally objected to their secretary fighting against their principles; the Unionists were for the most part shocked that a young man bred to the sea and of no political experience or maturity of thought should lightly rush in on what they held to be sacred ground, for in Scotland the solemnity of political responsibility is second only to that of religion. Hence a crisis arose which Shackleton's quick wit readily seized, and in formally announcing his intentions to the Council he offered to resign the secretaryship. This spirited action rallied his friends to him. Sir John Murray, whose strong, unconventional nature warmed to a man who had so much in common with himself when young, had been followed in the presidency by another who was less sympathetic; but the resignation was not accepted, and matters remained as they were for a time.

In February, Shackleton's eldest son was born, and the proud father on seeing the morsel of humanity only a few hours old exclaimed, " Good fists for fighting ! " Always fond of children, he overflowed with tenderness to his own, and little he cared for the critical attitude of his Council when his home held a wife and child that outweighed the whole world.

The months went on with many visits to Dundee and London, and much hard thinking as to future prospects in case of political success or failure. The dream of a short cut to fortune was still with him. He saw agents waxing fat on commissions earned, apparently, with ridiculous ease, and he tried his hand quietly at several agencies in which his name did not appear.

He acquired an interest in a tobacco concern, and when he announced at the Society that he had obtained a full-page advertisement of the Tabard cigarettes for the Magazine, only a very few and favoured friends knew who paid for it. His friend Mr. W. Beardmore held out hopes of profitable employment and a certainty of at least a financial equivalent of the secretaryship, so at last in July Shackleton resigned in earnest, was formally thanked by the Council, and at the earliest opportunity was elected a member of that august body.

This made a difference : it was now reasonable that he should desire to serve his country in Parliament, and the Unionists saw that even if he failed to get in he would be a credit to their cause. So the period of secretaryship came to a seemly end; the Society had five hundred more members than when he entered on his duties, and the advertisement account of the Magazine was proportionally even more enhanced. His chief assistant, . Mr George Walker, now assistant secretary of the Royal Scottish Geographical Society, always had the most loyal admiration for him, and has passed on to the present day many of the clever devices and attractive methods which Shackleton was always originating. This friend attended the candidate for Dundee on his week-end visits to that city and eagerly helped in the work of the campaign. He summed up his recollections thus· "What impressed me most were his optimism when circumstances were adverse, his vitality, physical and mental, and his geniality and kindness."

When free from the routine of his Edinburgh office Shackleton took a house at St Andrews for the summer, and he and his wife were devoted frequenters of the golf course, starting before the crowd for the first round, and making another in the early afternoon. Although Shackleton was the less expert of the two, and his play was such as not to win the respect of the caddies, he was possessed by the game and talked of nothing else at meals, at least his sister-in-law used to complain that this was so. One day he resolved to play an appropriate practical joke upon her, and conspired with the cook to produce an entrée of golf-balls beautifully concealed in egg and breadcrumbs and served on toast. The Shackletons enjoyed the struggles of their non-golfing relative to get her fork into the supposed rissoles, and the victim joined as heartily in the fun when she discovered the joke.

Dr. Charles Sarolea, Belgian Consul in Edinburgh and Professor of French at the University, was frequently of the party, and became engaged to Miss Julia Dorman, the heroine of the golf balls, before the sojourn was over. Shackleton and Dr. Sarolea proceeded to the Continent for a short visit in pursuit of the scheme for a great international news agency which still fluttered before him as an *ignis fatuus*. It brought him, however, an interesting interview with that sinister monarch, King Leopold II., the first of the many foreign royalties with whom he was destined to confer.

The idea of a new polar expedition was never very far from his mind, and in the course of this autumn he made ineffectual efforts to rouse enough interest in some wealthy people to loose their purse-strings. The political situation grew engrossing as the year went on, and Antarctic ambitions had faded for the time, when at last, on 4th December, the long-expected resignation of the Unionist Government took place. Canvassing for votes absorbed all his thoughts. Never was a candidate more welcome in the house of a voter, for to friend and foe alike he was light-hearted, cheery, and jocular, with a vein of ardent conviction when he touched on the wrongs of the British sailor that never failed to seize the attention of a maritime community. Once when in Edinburgh he missed a train, and rather than be late for a party meeting he took a special train, an impressive fact for a thrifty electorate.

The last few weeks before the election went by in a swirl of talk. There were serious meetings in great halls, where Shackleton was hampered by sharing the platform with his Conservative colleague, Mr. A. Duncan Smith, a young advocate born in Dundee who had stood for the city in the previous general election. Both had set forth the common principles of their allied parties—opposition to Home Rule for Ireland, fiscal reform in the direction of protection, a closer bond between the peoples within the Empire, and a strong Navy.

In his address to the electors Shackleton had given his adherence to all these, saying of Ireland :

" As an Irishman myself, in sympathy with the Irish people, I am convinced, moreover, that my country has been gradually emerging from sad and bitter times to a more peaceful and brighter era ; and that this has been accomplished by the wise measures, considerate administration, and great pecuniary assistance due to the Unionist Government."

As to fiscal policy :

"Actual free trade is an economic ideal, but as an industrial community we have to deal with the concrete conditions of the world of commerce and be prepared to use the weapons of the age. I go further, and say that we should be willing to consolidate our developing empire by a scheme of preference and fiscal union with a view to the ultimate realization of the British hope of free trade within the Empire."

These were the machine-made views ardently held by every Unionist candidate at the general election of 1906 ; but the real man shone out later in the address :

" I am whole-heartedly in favour of the manning of British ships by British seamen. I consider the employment of so many foreign seamen in our mercantile marine as not only an unsound position, but unfair to Britishers, and as a menace to the country in the eventuality of war with foreign Powers. The British merchant sailor is without an equal at sea. I speak of this from actual experience—from a working knowledge as a sailor trained in the hard school of the stormy seas throughout the world. I know the greater value of one British seaman in rough weather, as compared with three foreigners under the same conditions."

And in the final outburst :

" I appeal to you, workers, to return me—a worker myself, a worker by my hands as well as by my head, from the time I was a boy of sixteen—as one who by the hard road of manual toil has been brought up to understand the conditions of the working man and to sympathize with him."

The strong sincerity of this solidarity with labour was quickly recognized, and at one of the meetings a grey-haired working man shouted to him, "Come on our side, boy, and we'll put you in at the top of the poll "; and very likely they would have done it.

His opponents were Mr. Edmund Robertson (the sitting member) and Mr. Henry Robson (Liberals), and Mr. Wilkie (Labour). Mr. Robson had capped some sarcastic verses by Shackleton, who, at his next meeting, responded that he believed Mr. Robson had caught a disease and that he had got the infection of poetry from him; now he only wished that Mr. Robson would catch the salutary infection of fiscal reform from him as well. He went on to improvise a limerick on Free Trade, where the ghost of Cobden addressed Sir Henry Campbell-Bannerman :

> "Said Cobden, ' You simple C.-B.,
> I meant the world's trade to be free;
>   But when nations protect,
>   I should strongly object
> To allow them to fatten on me.' "

Such things provoked much laughter in the heated halls, and served to grease the wheels. Shackleton was at his best in dealing with the hecklers, who in Scotland have brought the practice of the cross-examination of candidates to a perfection unknown elsewhere. Here his quickness of repartee delighted his audiences, and he became as fast friends with his most pertinacious opponents as with his own supporters. It was a tough job sometimes, as when he supported Chinese labour in the Transvaal mines because a large majority of the people of the Transvaal wanted it; on which a heckler said: "So, since 95 per cent. of the people of Ireland want Home Rule you would give it ? " His reply was, " I'm an Irishman myself, and I would never give them anything that is not good for them." There was an analogous case in Scotland also which led to the question, "Do you approve of Poles being allowed to work in Lanarkshire mines ? " " I would shift every Pole," he replied.—" The South Pole ? " queried a voice — " The only poll I would not shift is that which I am to be at the head of—with Mr. Smith ! " Of Woman-suffrage he had never had a serious thought, but for the question flashed at him, "Would you give votes for women?" he had a happy inspiration. "Hush!" he said, in a hoarse whisper. "The fact is, my wife is present," and the roar of laughter solved the problem for the time.

He was most popular of all at the impromptu dinner-hour meetings at the various works or docks, where his knowledge of the sailor-man and his easy use of sailor-language made a delightful contrast to his auditors with the suave and formal Mr. Smith "I like that funny beggar!" shouted an excited worker after a typical yarn

When at last the polling day arrived, on 16th January, he was tired but undaunted, and the great voice which had roused the testy comment in the early days, "Dinna shout sae loud. Lord, man! we're no deef," had grown hoarse, as he said:

"This is my fifty-fifth and last meeting. Dundee will ever remain with me a memory of straightforward talk and straightforward answers on both sides."

While the votes were being counted his wife and her sister were waiting in their private sitting-room in a hotel late on a cold night. The fire had gone out, the window was open to catch the first sign of the result. Suddenly there was cheering in the distance, the sharp whirr of a telephone bell below, and on rushing to the stair with the question, "Who has got in?" the ladies met a little German waiter running up excited and hair on end calling out, "Robertson and Vilkee."

The former Liberal member and the Labour man had been returned with great majorities. Shackleton, who had nourished some hope of success up to the end, came in quietly, full of good humour, and that night enjoyed the best sleep he had had for months. He had done his best, and, like Alan Breck, he knew that he had been a bonny fighter.

The result of the poll was:

ROBERTSON (Liberal)	9276
WILKIE (Labour)	6709
ROBSON (Liberal)	5998
SHACKLETON (L.U.)	3865
SMITH (Conservative)	3183

SHORE JOBS. 1903–1906

No fewer than 769 had voted for Wilkie and Shackleton, a fact which shows how strong a hold he had on the working men. Many years later, when reminded at a lecture in Dundee of his immense popularity during this election, he remarked placidly, " Yes. I got all the applause, and the other fellows got all the votes."

Shackleton was much in London during February 1906 on one of the fortune-hunts into which his inexperience of business methods was constantly attracting him. This time it was a contract for the transport of Russian troops from Vladivostok to the Baltic by sea His associates had led him to look on this as almost a certainty, and he was preparing for a five months' absence to superintend the operations when the scheme fell through.

His friend Mr. W. Beardmore (now Lord Invernairn) found him a less sensational but more secure position in his great engineering works at Parkhead, Glasgow, where he acted as secretary of a technical committee which the firm had established for the investigation of gas-engines with a view to the more economical production of power. Here there were opportunities of meeting with hard-headed, practical men of business, shrewd in their outlook on commercial affairs, and alive to the possibilities presented by the application of the latest scientific research. It was a wholesome corrective to the excitement of visionary news agencies, and the hunt for contracts to yield fabulous profits.

Shackleton kept on his house in Edinburgh, leaving every morning at seven o'clock to take the train to Glasgow, and returning late in the evening; often not returning the same day, for there were frequent business visits to London, travelling by sleeper one night and returning in the same way the next.

In summer the Shackletons took a house at Queensferry almost under the Forth Bridge, and one day they saw a naval ship coming to her anchorage Shackleton got up on the roof of his house, attracted the attention of the officer of the watch, and signalled an invitation to the ward-room officers to come ashore and visit him. The ship was H.M.S. *Berwick*, and one of the officers who came ashore was Lieutenant J. B. Adams,

R N.R. The acquaintance ripened quickly, as is usual with naval men, and as the talk turned on exploration in the Antarctic, Adams said that if Shackleton ever went out again he would go with him. The idea of a new expedition, which was rooted in Shackleton's mind, had been peeping out at intervals for years past ; but something always arose to drive it back again into hiding. But the leaven was working, and the long equipment for his lifework was very nearly complete.

His childhood in Ireland, his boyhood at Sydenham, his hard-driven youth on the *Hoghton Tower* and the Shire Line tramps in all the seas, his post-graduate course in the easier and pleasanter routine of the Castle Line, his joyful preliminary dash into exploration on the *Discovery*, the refining influences of his marriage and the polished society of Edinburgh, the breaking in to business habits and the thrilling quests for political honours and speculative fortune, had all been fashioning and tempering his life as an instrument for great achievements. And all the time his friends were hoping and almost believing that he was "settling down."

An impression of the man at the close of his equipment and before he had entered on achievement, cannot be given better than in the words of Mrs. Hope Guthrie :

" I first met Ernest Shackleton early in 1904 at a friend's country house, where a small party had been gathered for the week-end. I remember that I hardly spoke to him until towards the end of the first evening, when three of us, including himself, being seated near the fire, one of us asked him some question about the Antarctic. The whole man flashed into an extraordinary vividness, he drew his chair eagerly forward and began to talk. . . . An hour later we awoke to the fact that here was a youthful and very modern mariner who had held us as spellbound as ever the immortal Ancient Mariner held the Wedding Guest, and that behind the eye that held us lay the dominance of genius

Next day, during a motoring excursion in the midst of a furious snowstorm, I struck another vein in the rich personality. The question, ' Do you love poetry ? ' was shouted at me, and above the wind I heard the rhythmic chant of great lines His love of poetry was passionate, and he stored his mind with —not passages—but whole poems and pages of Shakespeare,

Browning, the Bible, Kipling, Service, and others. Poetry was his other world, and he explored it as eagerly as he did the great Antarctic spaces. . . .

Everyone who knew him knew his love of Browning, and that he seemed to be an incarnation of that poet's virile faith and optimism; many also have marvelled at his memory, and indeed its scope and exactness were marvellous. I once asked him if, when he had read a page of Shakespeare in order to commit it to memory, he had ever to glance at it a second time, and he replied: "Sometimes to make sure of the punctuation!" Since his death the poignant memory abides with me of his repeating in a summer garden the beautiful dirge :

> 'Fear no more the heat o' the sun,
> Nor the furious winter's rages,
> Thou thy worldly task hast done,
> Home art gone and ta'en thy wages'

I sometimes wonder if the man who has a genius, not only for exploration, but for organization and leadership, does not occupy one of the most difficult and delicate positions open to genius. He is as absorbed in the plan of his work and in attaining the best tools and material for it as the artist or poet, but they may shut themselves away when the frenzy is upon them, and their seclusion is, upon the whole, respected; the explorer, on the contrary, must needs be in human contact upon every side and all the time, and it may work him not a little woe if in his fury of absorption he happens to hurt human feelings—or failings. Even the ordinary public man knows this difficulty.

I have heard Ernest Shackleton say with a sort of bewilderment after a spell of intense physical and mental exertion, 'You know so-and-so is hurt, *and I didn't mean it.*' And indeed I never knew any man greater hearted or more generous both in thought and deed. He was direct and simple as a child, with something often of the artlessness of a child, hated to refuse anything asked of him, and, when disillusioned and hurt himself, never 'nursed a grudge.' To many of the loyal friends who loved him he seemed even too generous.

When to genius is added a pronounced Celtic strain, the daring of the sailor, and eternal hopefulness, one may anticipate a certain amount of risk. A sailor is proverbially bad at 'driving a bargain,' and not a few harder beings have benefited by this knowledge. On every side his was 'the mould and mind to which adventures come.' . . His brilliant imagination and versatility were very attractively shot with boyishness, so that one could be happily sure of him treating 'daft' suggestions and romancings with proper attention and sympathy —as the ever young ever do . . And as he was the ready comrade in any frolic or flight of fancy, so we found him always the true comrade in trouble and sorrow ; added to his magnetic qualities and the tonic of his personality was a tact and delicate sympathy in which one could unfailingly confide. He invigorated and inspired one to fight the good fight, not only in life's big things, but also in ' the littlenesses that hold so much of the great woeful heart of things ', all these he understood, and heartened one for the fray or for endurance."

BOOK II
ACHIEVEMENT

". . Know, not for knowing's sake,
But to become a star to men for ever,
Know for the gain it gets, the praise it brings,
The wonder it inspires, the love it breeds,
Look one step onward, and secure that step."
                                    ROBERT BROWNING.

Ernest Henry Shackleton, aged 33.

CHAPTER I

SHACKLETON ASPIRES. 1906–1907

"Are there not, Festus, are there not, dear Michal,
Two points in the adventure of a diver?
One when, a beggar, he prepares to plunge;
One when, a prince, he rises with his pearl?
Festus, I plunge!"

ROBERT BROWNING.

THE plans of renewed exploration which had been maturing in Shackleton's mind were kept to himself during the later months of 1906, in order not to add to his wife's anxieties; and perhaps his first clear hint was given in a letter to a friend, dated from Parkhead Works, Glasgow, on 26th December, the day when a sudden blizzard swept the British Isles and laid all England under a blanket of snow.

"You can imagine how work presses on me here when I tell you that even on Christmas Day I had to make my usual daily journey from Edinburgh to Glasgow.
"You will be pleased to hear that the most interesting Christmas present I got was from my wife, who on Sunday morning introduced to the family a splendid little girl.
"I see nothing of the old *Discovery* people at all. We are all scattered, and the fickle public are tired of the polar work at present. What would I not give to be out there again doing the job, and this time really on the road to the Pole!"

A week or so later he opened his mind to Mr. Beardmore, who, attracted by the idea and full of confidence in the promoter, agreed to guarantee a considerable part of the estimated expense of an expedition, and the spirit of Shackleton leaped before him to draw up detailed plans for his prospective triumph.

104 THE LIFE OF SIR ERNEST SHACKLETON

It was easier to plan an expedition than to tell his wife, not yet recovered, that he had renounced the last of his shore jobs; but he did it very humorously, pointing out how delightful it would be for him to return to the new baby with stories of how he had climbed the border mountains of Victoria Land and looked over into green valleys where quaint little people were carrying on a fairy-like existence in conditions such as never were in the humdrum world. The announcement, tenderly put, was bravely taken, and heart-stricken as she was with the thought of a long separation, Mrs. Shackleton offered no discouragement, but set herself to help forward the plans as they developed.

It was not lightly that Shackleton undertook to equip an expedition on credit, for it amounted to this, as he pledged himself to redeem the guarantees from the sale of his prospective book and the results of his future lectures if the venture proved successful. He was much disquieted by the fear that a foreign expedition would set out for Victoria Land before he could get ready. On 4th February he wrote to a friend:

"*Re* the Expedition: I have been full of anxiety about the other nations, and am hoping that by the end of this week I will have nearly all the money guaranteed and shall announce it before the 12th. I have had three years' struggle and tackled seventy-odd rich men, but I think the end is in sight; only one cannot say it until the money is actually safe. You will know before any one else in London.

"I think that if the other nations can get money to go we ought to, but I have put a black mark against a good many men!"

The seventy-odd rich men have purchased a large section of oblivion for themselves by their caution, and no geographer will seek to penetrate the shade that hides them. A few generous supporters were found who gave with gladness what they could afford, and if the limitation of their resources made it impossible to have a really good ship for the expedition, that only enhanced, as it happened, the reputation of the good workman who could use the inferior tools. Amongst the willing supporters we may mention the Misses Dawson Lambton,

SHACKLETON ASPIRES. 1906-1907

who had presented the balloon to the *Discovery*, and had been devoted friends of Shackleton from the day when he showed them over that ship in 1901; Mr. G. A. McLean Buckley (New Zealand), Mr. Campbell Mackellar, Mr. Sydney Lysaght, Mr. A. M. Fry, Colonel Alexander Davis, Mr. H. H. Bartlett, Mr. Wm. Bell, Shackleton's kinsman, and in a very special way his brother-in-law, Mr. C. H. Dorman.

Fear of foreign competition was a sharp spur to hasten preparations and allow the British Antarctic Expedition, 1907, to be first in the field for the new race for polar honours. Dr. Arçtowski, a prominent member of the *Belgica* Expedition, was struggling, vainly as it proved, though then with every prospect of success, to lead a new Belgian venture; and Dr. Jean Charcot was building a polar ship with the hopeful name of *Pourquoi Pas?* to carry the French flag again into the Bellingshausen Sea. Other foreign projects were on foot; but except for a scheme by his old friend, Michael Barne, to explore Grahamland, and Bruce's hopes for the resumption of research in the Weddell Sea, Shackleton knew of no projects which were likely to bring the British flag again into the Antarctic.

On 11th February Shackleton and Arçtowski were guests of the Kosmos Dining Club of the Royal Geographical Society, and each of them outlined his plans to a small company. Next morning Shackleton's expedition was announced in the newspapers, and a few weeks later the details were printed in the *Geographical Journal*, the organ of the Royal Geographical Society.

The plan was that from nine to twelve men should be landed in February 1908 at the old winter quarters of the *Discovery*; the ship should then go back to New Zealand and return early in 1909 to bring them home. Three parties were to be formed, one to travel eastward over the Barrier and explore King Edward Land; one to winter near Mount Melbourne, and go westward across the mountains to the South Magnetic Pole; and one, the main party, to go south on the Barrier, keeping farther away from the mountains than Captain Scott had done, to make a determined effort to reach the South Pole.

Transport was to be effected by the use of Siberian or Manchurian ponies instead of dogs, and a specially-adapted motor-car was also to be taken ; but the sledges were to be of such a size that they could, in case of necessity, be hauled by men, while depots of provisions were to be left at each 100 miles, to be picked up on returning along the outward route. A kinematograph and phonograph were to be taken for recording the ways of penguins, and various modifications of dress and diet suggested by previous experience were to be adopted. Research in magnetism, meteorology, geology, and biology was to be pursued, and the plan of the expedition provided for the working up of the results by the specialists themselves on their return.

The plan was clear, concise, and scientific without any effort at being sensational, and, save for the probable benefits to weather forecasting in the southern hemisphere, and for the value of improved variation charts to navigation, there was no suggestion of a utilitarian appeal.

The announcement brought an immediate flood of applications for positions on the staff, but, before considering any of these, Shackleton invited a number of the old *Discovery* party to join him. First of all came Dr. E. A. Wilson, the friend of all in the *Discovery's* ward-room The offer was made with touching affection and generosity ; but Wilson had undertaken an important research into grouse disease, the completion of which demanded still two years of work, and he felt that honour compelled him to stick to his job. There was a hailstorm of letters and telegrams between the two, and Shackleton used every argument to secure his friend, whose final reply concluded :

"I can only tell you that I would have jumped at your offer had I been free, and we would have had a really good time out there together As it is, you will find some one else, and I shall hope and pray that you will make a really thorough and successful job of it to make up for all your former disappointment. Bless you, my dear old boy, I can tell you honestly that I value your friendship above all things You are a real trump to write as you have.—Yours ever,   BILL."

SHACKLETON ASPIRES. 1906-1907

A visit to Plymouth found Skelton reabsorbed into the Navy, and contented to stay there, and Hodgson immersed in his marine zoology by the shore, and unwilling to face the Antarctic again. An offer of a place on the expedition to Mulock (who had succeeded Shackleton on the *Discovery*) brought the puzzling answer that he had already volunteered for Scott's expedition; and a few days afterwards, on 24th February, a belated reply from Michael Barne, to whom Shackleton had offered the leadership of his Magnetic Pole Party, brought the staggering news that he had been acting for some time as a confidential agent for Captain Scott, who contemplated a new expedition from his old base in the course of a year or so. A few days later a letter from Captain Scott, then in command of H.M.S. *Albemarle* at Gibraltar, confirmed the news, and told of his surprise and concern at the announcement of Shackleton's plans in *The Times* of 12th April. Shackleton replied, expressing his concern on learning of Scott's project, for which he was quite unprepared.

The correspondence which followed between the two and Dr Wilson, whose regard for both was equal, reflects the greatest credit on the good temper and restraint of all three It was friendly, temperate, and restrained, though tense with the emotion of the matured ambition of two lives on the point of collision. Captain Scott asked Shackleton to modify his plans by adopting a base other than the old *Discovery* winter-quarters from which he himself hoped to lead a new expedition. Shackleton was most reluctant to abandon his scheme; but as in many crises that were to come, the spirit of his Quaker ancestors calmed the storm that was rising in his mind, and on 6th March he cabled to Captain Scott :

"Will meet your wishes regarding base. Please keep absolutely private at present, as certain supporters must be brought round to the new position. Please await second letter."

That fight was won, though Shackleton believed that by altering his plans he not only reduced his chance of getting to the Pole, but introduced the necessity for a larger and more costly ship than the funds in sight could afford. Nevertheless out of respect for his old leader he wrote to Captain Scott agreeing to abandon his intention of making M'Murdo Sound, or the Barrier west of 170° W , a base for his attempt on the Pole, which he decided to approach from King Edward Land.

The ground thus cleared, Shackleton proceeded to select his companions. He was determined to be hampered by no committee or external authority of any kind ; the whole responsibility rested with himself Yet he sought advice from all experienced persons ; he saw much of Dr. Nansen, who was then the Norwegian Minister in London ; he met Roald Amundsen just returned from the first voyage through the North-West Passage ; he took counsel with all who had experience in the outfitting of expeditions with gear and clothes and food, and he consulted the men of science most skilled in all the branches of knowledge which his expedition was likely to advance. He kept the promise made to Lieutenant J. B. Adams, R.N R., at Queensferry a year before, and appointed him as meteorologist ; and from the crowd of volunteers ultimately selected Mr. Raymond Priestley as geologist; Sir Philip Brocklehurst, Bt., who contributed to the expenses of the expedition, as junior geologist ; Mr. James Murray, who had been assisting Sir John Murray in his survey of the Scottish lochs, as biologist ; Dr. E. Marshall and Dr. A. F. Mackay as surgeons, the former acting also as cartographer, the latter as zoologist ; Mr. G E. Marston as artist ; two old *Discovery* sailors, Petty Officer Ernest Joyce and Frank Wild, for positions of general utility; and a reinforcement of scientific men was expected from Australia.

An office in London was secured at 9 Regent Street, Waterloo Place, and a manager was appointed. With him Shackleton paid a short visit to Norway in the end of April to purchase sledges, furs for boots and mittens, sleeping bags of reindeer skin, ski and similar equipment for which Norwegian firms were the best in the world. In Norway he inspected a polar ship, the *Bjorn*, which had just been built, a fine vessel of 700 tons with powerful triple-expansion engines, the very ideal vessel for his purpose, for her strength and speed would afford the best chance of reaching the hitherto inaccessible base which his loyalty to the wishes of his old commander had led him to select for his expedition. The price was necessarily high, and as the funds were small, Shackleton had no alternative but to shoulder the burden of looking for some cheap old craft.

He heard of one in Newfoundland, a sealer of little more than 200 tons net, built in Norway, which for forty years had been justifying her name of *Nimrod* by hunting seals in the Arctic Seas. An agent in Newfoundland was instructed to inspect and survey her, and the report being satisfactory, she was bought in May, and on 15th June she appeared in the Thames, an ice-worn hull, rigged as a schooner with an engine incapable of driving her at more than six knots, and filthy holds stinking of generations of putrefying seal-oil. Shackleton had everything arranged for reconditioning her. He knew the hull to be sound and fit to battle with ice and storms, and it was not long before the skilled hands in Green's famous shipyard at Blackwall had made her all that his foresight pictured her. The old masts and rigging were cleared out; the holds scraped and cleansed, new accommodation for the scientific staff contrived out of part of the after-hold, and three new masts set up, altering her rig from a schooner to a barquentine.

Meanwhile the Shackletons had left Edinburgh for the time and took a furnished house for a month or two at Palace Court, Bayswater Road. From his Regent Street office Shackleton controlled an immense organization. He had agents in Manchuria buying ponies and arranging for their shipment to New Zealand; agents in New Zealand buying dogs in Stewart Island, descended from those of Borchgrevink's expedition of ten years before; agents in Australia arranging for scientific reinforcements; agents in Norway seeing after the equipment, while he himself was everywhere in England ordering, selecting, inspecting food-stuffs, seeing to the building of the hut which was to be taken out in sections to form the winter quarters of the expedition; watching experiments with motor-cars, settling difficulties with the staff, engaging new men, arranging for officers and crew for the ship. He was always increasing the number of the "seventy-odd rich men" whose firm hand on their money-bags would have strangled the enterprise of a man who lived more in the world of facts and less in those regions of the imagination where Shackleton, though still only a beggar preparing to plunge, saw himself surely rising as a prince.

By the beginning of July the preparations were well advanced ; the Admiralty and the Royal Geographical Society provided charts and instruments for navigation and magnetic work, many prominent firms made contributions in kind, and an exhibition of the equipment was held at 9 Regent Street, where a constant stream of visitors inspected sledges, tents, clothing, cooking apparatus, foods, and instruments. A special feature was the large supply of dried milk taken, and the plasmon biscuits for sledge journeys Shackleton's long experience in handling stores and stowing cargo led him to make one innovation which proved to be of great utility. Instead of packing each commodity in the usual boxes familiar to the various trades concerned, he designed a packing-case of "venesta" boards. This material is a composite board made up of three very thin layers cemented together. It is very light, very strong and weather-proof, and 2500 cases were made of the uniform dimensions 30 inches by 15 They packed close for stowing, and when emptied the boxes were useful for building partitions in the hut and for many other purposes. As compared with ordinary boxes, they ensured a saving in weight of more than 4 tons, so allowing additional stores to that amount to be carried. Everything was got ready in an incredibly short time, and on 30th July the *Nimrod*, with all her cargo on board, steamed down the Thames.

After the ship had started a message was received from the King intimating his intention of inspecting the vessel at Cowes, and she was brought into the Solent. Here, on 4th August, almost exactly six years after the *Discovery* had received a like honour, the *Nimrod* was visited by King Edward and Queen Alexandra, accompanied by the Prince of Wales, Princess Victoria, Prince Edward, and the Duke of Connaught. The visit was as informal and heartening as that of 1901, and to complete the resemblance His Majesty invested the leader with the 4th class of the Victorian Order, while Queen Alexandra presented a Union Jack to be hoisted at the farthest south—

at the Pole itself, Shackleton confidently believed, as he proudly accepted it.

On her way down Channel the *Nimrod* called in at Torquay, where the Shackletons were residing at the time; and on 7th August the ship started on her three and a half months' voyage to New Zealand, under the command of Captain R. G England, who had been first officer of the *Morning*, with John King Davis, Æneas A. Mackintosh, and A. E. Harbord as subordinate officers. Some of the scientific staff sailed with the ship; but there was much organizing work still to be done by the leader in England.

In the September issue of the *Geographical Journal* Shackleton for the first time made a formal announcement of the change of plan arrived at in the spring in deference to Captain Scott, although, as the latter was not yet ready to publish his own plans, no reason for the change could be given. It was simply stated that the base of the expedition would be in King Edward Land instead of being at the old winter quarters of the *Discovery*; that three exploring parties were to be sent out from the base, one due south towards the Pole, one south-eastward to explore King Edward Land, and one eastward, along the coast of that region, all three leading into parts of the Antarctic which were totally unknown.

Shackleton had nearly two months of exhausting work at home completing the scientific staff, hurrying up delayed equipment, smoothing out hitches in the preparations all over the world, and making arrangements with publishers, newspapers, and lecture agents for the publication of the results of the expedition on his return If successful he was confident of funds enough to pay off all expenses and leave a substantial balance on which he and his family could live in affluence ever after. Through all the rush of the final weeks he was harassed by other cares, but at last the time came for his departure. On the last day of October he left England, his last sight being the figure of his wife as she stood watching on Dover Pier. Her image remained with him on his rush through France and on all the voyage through the Mediterranean and the Red Sea as it had done on the *Flintshire* ten years before. And as in all his leave-takings, he suffered from home-sickness. He wrote to his wife :

"Honestly and truly, parting from you was the worst heart-aching moment of my life. If I failed to get to the Pole and was within 10 miles and had to turn back it would or will not mean so much sadness as was compressed into those few minutes."

And again :

"I can never, never put into words all you meant to me that day standing on Dover Pier, and all it has meant to me ever since . . . for you care for me to the height of letting me go to fulfil my destiny."

But, as ten years before, he did not let others see him as a sentimental dreamer. He took his part in the social life on board the *India*, where he met a new friend, Mr. Sydney Lysaght, as a fellow-passenger. He lectured on his expedition to the gratification of all who heard him, and when he landed in Melbourne, in time to give a great public lecture on 3rd December, he had recovered from his fatigue and was ready to face the crowded weeks that followed.

He spoke to a large audience in Sydney on the 6th, met Professor Edgeworth David, who was to accompany him to King Edward Land as an adviser in geology, and return with the *Nimrod*, and Douglas Mawson, who was joining the expedition as a physicist. Then a few days of rest in the crossing to Wellington, and more lectures there and at Christchurch ; all of them largely attended and each bringing in several hundred pounds. This money was greatly needed for the expedition, but with his usual quixotic generosity, Shackleton devoted all the proceeds to local charities. No wonder that when, on the last day of the year all was ready, he found the whole Commonwealth of Australia and the Dominion of New Zealand roused to enthusiasm, the governments made generous contributions to his funds, the warm-hearted people overwhelmed him with contributions in kind, and he responded to it all like the war-horse smelling battle from afar. It was indeed a year to look back upon with pride; from the beginning of January, when his plans were first unfolded in Scotland, to the end of December, when his ship lay deeply loaded in Lyttelton Harbour, his companions gathered around him, dogs and ponies brought from the ends of the Earth, and all the units of the expedition assembled at the appointed date, none lacking, all efficient and everything called into being by the dominance of his own individual will But he could not do impossibilities; stow as closely as he could, the little ship could not take the full equipment, and because of the lack of money to buy the *Bjorn* he had to leave behind several of the ponies and much of the material which, as the event proved, might have made all the difference and carried him through to the Pole.

No one whose ambitions have not been defeated by lack of money at the critical moment when the hour and the man are ready for great deeds, can realize to the full the bitter irony of the distribution of wealth in hands whose controlling head, with all its powers of acquisition, lacks the divine instincts of insight and generosity; but Shackleton was not the man to cloud his gratitude to those who helped by any such reflections.

CHAPTER II

THE *NIMROD*. 1908

"We live in deeds, not years, in thoughts, not breaths;
In feelings, not in figures on a dial.
We should count time by heart-throbs"

P. J. BAILEY.

THE little *Nimrod* had proved herself a stout ship on the voyage out from England; but she was slow under steam and made very poor progress under sail. It was necessary to economize the small stock of coal she could carry, and Shackleton decided that she must be towed all the way to the Antarctic Circle. The New Zealand Government contributed half the cost of the tow, and Sir James Mills, chairman of the Union Steamship Company, made up the rest. The idea was a daring one, for never before had any one attempted to tow a vessel hundreds of miles away from port in the stormiest part of the Southern Ocean. To a young people the unprecedented always appeals, and all New Zealand warmed to the adventure. The vessel appointed for the task was the *Koonya*, a steel steamer of over 1000 tons, under the command of Captain F. P. Evans.

On New Year's Day 1908 the expedition received a great send-off from Lyttelton Harbour, and Shackleton found himself at last on the deck of his own ship with all the worries of his year of scheming, begging, and bargaining behind him and a bright prospect ahead. The *Nimrod* was terribly overcrowded with her own company and the landing party, and Professor David, the patriarch of the crowd—he was in his fiftieth year—as a passenger to the Barrier. Nevertheless another passenger was squeezed in, a keen yachtsman, Mr. George Buckley, who, after doing a great deal for the expedition in New Zealand,

THE *NIMROD*. 1908

could not resist the temptation of a run to the Antarctic Circle, and made up his mind to go just two hours before the start. The *Nimrod* was towed by a long steel rope shackled on to her two chain cables, which were run out one on each side of the bow, almost to their full length, thus lightening the bow of the ship and forming a sort of shock-absorber for the jerks of the tow-rope.

The sea was not kind, and very soon the waves were breaking over the *Nimrod's* forecastle, deluging the ten-stalled stable in which the ponies were tethered, and the nine dogs which were chained out of reach of each other wherever space could be found. On the fourth day a carrier pigeon was released, carrying a message to New Zealand; and as the day went on the storm increased, the ship rolling fifty degrees on each side of the perpendicular, so that it was almost impossible to walk along the deck. For ten days a gale blew, rising sometimes to such terrific force that the *Koonya* had to stop, and both ships were laid to, wallowing in the tremendous seas. The scientific staff were organized into a stable-watch, and never for a moment, night or day, were the ponies left untended, all efforts being made to keep them on their legs. One poor beast fell in its narrow stall, and was turned over on its back by a roll of the ship. Efforts to get it round again were unavailing, and at last it had to be shot. The bulwarks were smashed by a great wave, and some damage done on deck. Cooking was almost impossible, and there was not a dry spot on the ship, nor for a fortnight could any one have dry clothes or a dry bed. Through it all the *Koonya* forced her way southwards, and on the 13th the wind abated, and at last it was possible to sort things up a bit.

Next day the first iceberg was sighted, and on 14th January the Antarctic Circle was reached, and here in perpetual daylight the long tow of 1500 miles ended. A number of sheep had been carried on the *Koonya*, and these were now slaughtered and the carcases hauled on board the *Nimrod* by means of a rope. The first half of the consignment arrived safely, but the second lot was lost through the parting of the rope between the tossing ships.

Before leaving New Zealand Shackleton had been sworn-in as a postmaster, and supplied by the New Zealand Government with a special issue of stamps for use on the expedition while in the Antarctic. These were the ordinary red penny stamp of New Zealand overprinted in green with the words " King Edward VII Land," and there was a complete postal equipment of dating stamps, registered letter labels, and the proper books for keeping an account of Post Office business. The stamps were a genuine issue, and now have considerable philatelic value. The total number of stamps printed was 24,000 of the face value of £100, and they were all carried on the expedition, with the exception of 448 supplied to the offices of the Universal Postal Union, and 60 retained as a specimen sheet by the New Zealand Government. These stamps were first used for the mail carried by Captain England to the *Koonya*, when he put Mr. Buckley on board that vessel to return to New Zealand on 15th January 1908 Captain F. P. Evans deserves the greatest credit for the splendid feat of towing the *Nimrod* safely through terrible weather, and for bringing his ship to the Antarctic Circle, which she was the first steel-built vessel to cross.

The *Nimrod* was now alone under her own steam with a heavy task before her, and at first she enjoyed the best of good luck, for she steamed through a labyrinth of huge floating icebergs without difficulty or delay, and entered the ice-free waters of Ross Sea without meeting any pack-ice at all. Every previous expedition had been held up for from one to six weeks by heavy pack-ice, through which it was very hard to force a way. On 23rd January the Great Ice Barrier was sighted in longitude 172° W.: it had never been reached before in such a short time by any expedition. To most of those on board the sight was new, and as the great ice cliffs grew to their full majesty as the little ship ran up close to them, the stupendous magnitude of the Antarctic rose upon their minds, for this was but a point on the edge of that three-hundred-mile-long floating wall, which bounded unknown hundreds of miles of nearly level surface stretching perhaps to the Pole itself, or even right across to the Weddell Sea. It was the period

THE *NIMROD* IN ROSS SEA.

of exhilaration in the psychology of polar travel when the heart of the explorer rejoices in the calm weather after stormy weeks, in the unsetting sun, the unbroken blue of sky and sea, and the white Barrier veiling the field where glorious deeds were about to be done.

Shackleton had in an exceptional degree the happy power of delegating responsibility and of refraining from interference when he had done so, though ever watchful to resume control should his helpers break down. On the way south he had been occupied mainly in the study of the land party, now brought together for the first time. The conditions were such as to reveal any bodily or temperamental weakness that might exist. He soon found that there was a tendency for groups to be formed, the Australians, for instance, holding together on the one side and the English members on the other, although the genial personality of Professor David did much to draw all together. Shackleton would not have hesitated to send back any man who developed ill-natured qualities; but he found them all good, noting merely the friendships that sprang up, or became obvious, so that he could group them into congenial parties.

From the longitude at which they had reached the Barrier Shackleton knew that he was close to the inlet where Borchgrevink had landed, and where the *Discovery* had lain alongside the low ice-wall to send her balloon ashore; but the ice cliffs now were high and changed. Slowly it was borne in upon him that a tremendous change had taken place; miles of the old Barrier front had been broken away to a depth of many hundreds of yards; and, in place of the inlet in which the *Discovery* had lain so snugly, there was now a great wide bay, a very playground of whales, which were spouting on all sides. Shackleton called it the Bay of Whales. It was here that he had decided to make his winter quarters had it retained its old outline; but now he felt that it would not be a reasonable risk to establish his party on the floating Barrier, which he foresaw might quite possibly break away again and drift with the whole expedition to certain doom. But, even if he had resolved to take the risk of landing there, as Amundsen did four years later with complete success, he could not have done so. As the *Nimrod* was coasting the Barrier in clear water a line of pack-ice mixed with icebergs began to appear on the northern horizon, and a northerly breeze was driving the pack in on the Barrier, narrowing the track of clear water and threatening to fill the Bay of Whales with ice.

He took an instant decision to abandon the attempt to land on the Barrier, for landing was not the matter of a moment, but a process requiring many laborious days of fine weather. Instead, he resolved to push on eastward to King Edward Land, where he hoped to find some spot within reach of solid ground where a winter base could be placed securely. The *Nimrod* got out of the Bay of Whales just in time, and set her course to work round the northern edge of the drifting pack, and so eastward towards the land. The motor-car, which had been unlashed and was standing on deck under the derrick ready to be swung on to the ice, was made fast again, and the ship's company faced the new move in the highest spirits. But neither Shackleton nor England, on whom lay the burden of the safety of the expedition and of the ship, viewed the situation or the prospect with elation. Again and again they put the ship's head into some promising eastward lead, again and again they were forced northward to avoid besetment. The weather grew worse, fog came on, all prudence—and Captain England was essentially prudent—cried for a return westward; but Shackleton, bound by his promise to avoid M'Murdo Sound, pressed on towards the inaccessible land, until the growing danger of the destruction of the expedition forced him to desist. "It must be part of my life," he wrote, "that I go on striving for the things that are out of reach."

That the decision was no light matter, but one which counted in heart-throbs, meant an ageing beyond the power of years, is apparent from the letter to his wife, which he sat up all night writing. It was dated 26th January 1908, and we quote the larger part·

"What a difference a few short hours can make in one's life and work and destiny I have been through a sort of Hell since the 23rd, and I cannot even now realize that I am on my way back to M'Murdo Sound, and that all idea of wintering on the Barrier or at King Edward VII. Land is at an end; that I have to break my word to Scott, and go back to the old base, and that all my plans and ideas have now to be changed, and changed by the overwhelming forces of Nature . . . All the anxiety that I have been feeling coupled with the desire to really do the right thing has made me older than I can ever say. . . . I must now write my heart out, and it is to you alone that I can do so, for I never, never knew what it was to make such a decision as the one I was forced to make last night. . . . You can realize what it has been to me to stand up on the bridge in those snow squalls and decide whether to go on or turn back, my whole heart crying out for me to go on; and the feeling against, of the lives and families of the forty odd men on board. I swept away from my thoughts the question of the Pole, of the success of the expedition at that moment, though, in doing so I know now in my calmer moments that I was wrong even to do that; that the money was given for me to reach the Pole, not to just play with, according to my ideas of right and wrong; that I had a great public trust which I could not betray; that in all ways my one line of action should have been the one I am taking now: but that was not what weighed with me then, and I feel that you will understand me in it all . . . To the north close in the offing lay the heavy pack, and I saw no chance of going north and trying again from the northwards of the pack a way to the east. My heart was heavy within me, for here was a direct check to my plans; if I had not promised Scott that I would not use "his" place, I would then have gone on to M'Murdo Sound with a light heart; but I had promised, and I felt each mile that I went to the west was a horror to me. I could see nothing for it but to go there, for after the fact became apparent that Balloon Inlet had gone, then obviously any idea of wintering on any other part or inlet of the Barrier would be suicidal and fraught with most serious danger, not only to the success of the expedition, but also to the lives of all the men whom I am responsible for. My promise was the one thing that weighed in the balance against my going back at once, and I gladly saw towards 6 p.m. a loosening of the pack to the northward, and after a long talk with England, who put the seriousness of my position frankly before me by the attempt—the shortage of coal, which even then was only sufficient to ensure the arrival of the ship at New Zealand; the strained condition of the vessel; the fact that even if we eventually arrived at King Edward VII. Land I might not be able to find a safe place to discharge, and would probably have to abandon it in view of the enormous masses of land ice and hummocked-up pack that was breaking away, which would make the ship's position untenable, my duty to the country and King, since I was given the flag for the Pole; and lastly, but not least, my duty to all who entrusted themselves to my keeping. I myself recognized the weight and truth of all he said, and I knew at the same time that he was heart and soul with me, and had no thought of his personal safety, indeed neither of us as we stood on the bridge the same morning thought of that, but of the safety of the laughing, careless crowd of men who little thought or dreamt what our feelings were as the ice was closing in, to them it was merely an interesting episode, and gave them an opportunity of a nearer view of seals and penguins. . I was determined in my mind that if difficulties were increased, then all that was to be done in the circumstances was to increase the labour so that a successful conclusion should attend our efforts.

I said to England that as regards the fuel question which was so urgent a one, that I was willing for him not only to burn all easily available woodwork on the ship, but also to burn the deck-house, cut away the mizzen-mast, burn it and the main top-mast, anything at all that would further our object and gain time, so that I could carry out my personal promise to Scott, this weighed with me even more than I ought to have allowed it to when I come to think of it in calmer moments now, but I felt that I could not turn back without trying from the northward a bit, so I told England that if I could not get east within forty-eight hours I would turn back to M'Murdo Sound as there was no other place I could go to and make the expedition a success. . . . I realized that to push farther on then would be madness, we could not lie where we were on the chance of it clearing up, for it might be days and our precious stock of coal would dwindle away, the ship will not sail and must depend on coal alone, so with a heavy heart I gave orders for turning back. All that night from 8 p.m when we turned till midnight when we ran or rather thumped into a rising sea and had the pack to the north of us, and so has ended my hope of reaching King Edward VII Land. My conscience is clear, but my heart is sore, and writing now I feel it as much; but I have one comfort, that I did my best; if I had gone back without risking and trying all I did, and if eventually I got the Pole from M'Murdo Sound base, it would have been ever tarnished and as ashes to me, but now I *have* done my best, and if the whole world were to cry out at me, which I am sure they would not, even then I would not worry myself, for I know in my own heart that I am right. . . . I have now put down all that has been on my mind, and I know that you will know that I have done all that any man could have done under the circumstances."

The battle had been fought out, and though the storm in Shackleton's mind had hardly subsided in the three days that brought the *Nimrod* to the base of Mt. Erebus, he put the past from him and threw his whole tremendous energy into safe-guarding the future. M'Murdo Sound was frozen over for more than 20 miles northward from the *Discovery's* winter quarters, and Shackleton waited for five days at the edge of the ice, hoping that it would break up and go out. An unhappy accident to Mackintosh, the second officer of the *Nimrod*, who was to have joined the land party, resulted in the loss of an eye and deprived the expedition of the services of a singularly active and fearless man. Captain England was fretting at the delay, and he had suffered more than any one else from the heavy strain of responsibility in difficult circumstances. On 3rd February Shackleton changed his plans once more, abandoned the hope of reaching the old *Discovery* base, and commenced to land the stores at Cape Royds on Ross Island immediately under Mt. Erebus. A good site was found for the hut on a flat peninsula, and although there was still a broad stretch of sea-ice between the ship and the shore the landing began, stores being sledged across. The motor-car was got out on the ice, and after a time it did some useful work, though much could not be expected from it on sea-ice always liable to break up. The ponies were got ashore with difficulty, as they were in poor condition after their five weeks at sea. One was so bad that it had to be shot.

Day by day the ice broke away more and more, and the ship came nearer in; but as 180 tons had to be landed and carried, progress was slow, and when the hideous snow blizzards of the region descended on them the men had some bitter experiences in saving cases and ponies on the splitting and driving ice. All hands worked like heroes, the leader most of all, and if any had a touch of the slacker in him he dared not show it when Shackleton was in sight. During all that strenuous time Shackleton suppressed his own feelings; he had a smile and a cheery word for all, and the impression which those days left on the men who served with him is well put in the feeling words of Dunlop, the engineer of the *Nimrod* in a private letter to a friend: " He is a marvellous man, and I would follow him anywhere "

The extreme caution of Captain England in approaching the ice and his anxiety to take the ship to sea on any change of weather for the worse, were hampering the work of landing stores and reducing the coal supply by unnecessary steaming. The other officers were indignant but obeyed; the shore party were less restrained in their opinions, and only Shackleton, who felt the delays most of all, upheld the authority of the captain, reprimanding one important member of the expedition who had made a disparaging remark about England in his hearing, and saying that he would not tolerate such language about his second in command, and would send any man who dared to repeat it home again in the *Nimrod*. More than once the captain's caution had threatened the safety of the wintering party, and most reluctantly Shackleton had been driven to the conclusion that England's health had suffered so seriously as to impair his usefulness. Still, he resolved to uphold the captain's authority until the ship reached New Zealand, and to leave it to Mr J. J Kinsey, the agent and friend of the expedition, to decide under what commander the *Nimrod* should be sent back the next year. On the critical question of the coal supply Shackleton had to exert all his powers of reason and persuasion, and even to appeal to his authority as leader of the expedition before England would abate his demand for 100 tons to be left in the bunkers in order to carry the ship back to port. The leader went, indeed, to the extreme limit of concession in reducing the amount to be landed for a year's supply for the winter hut from 30 tons, the quantity set out in the plans, to 18 tons; but he did so and left 92 tons on board the *Nimrod*. This, as it happened, was 30 tons more than was required on the voyage to Lyttelton; but then, as always, Shackleton took the greater risk for his own party.

THE *NIMROD*. 1908

Professor Edgeworth David had been persuaded to cast in his lot with the shore party, and proved to be not only a wise adviser on scientific matters, but a soothing influence whenever friction seemed about to appear between the young men. The vexatious delays, the incessant hard work, the want of sleep, and constant anxieties arising from sudden storms and the insidious growth of new ice, had brought the nerves of every one on shore and on board almost to the breaking point by the time that the last ton of coal was landed. Then, as the last boat got back to the ship and was hoisted in at 10 p.m. on 22nd February, Shackleton and his fourteen companions bade farewell to the *Nimrod* · she proceeded on her way to warmth and safety, they turned to the chilly task of setting their house in order and preparing for their great adventure.

Shackleton was fortunately unaware of an incident which followed the return of the ship to Lyttelton. Some discontented or mischievous member of the crew seems to have spun a yarn to a too credulous reporter, as the result of which a newspaper published a story of an altercation between the leader and the captain which it was alleged ended in a struggle. Captain England immediately published an emphatic denial, and all the officers and most of the crew signed a letter to Mrs. Shackleton expressing their disgust at the false report and their experience of the constant courtesy and consideration of their leader.

The sketch map on p 124 shows the track of the *Nimrod* from the Antarctic Circle to and from Cape Royds. It also shows the route of the Southern journey described in the following chapter, and that of Professor David's journey to the Magnetic Pole.

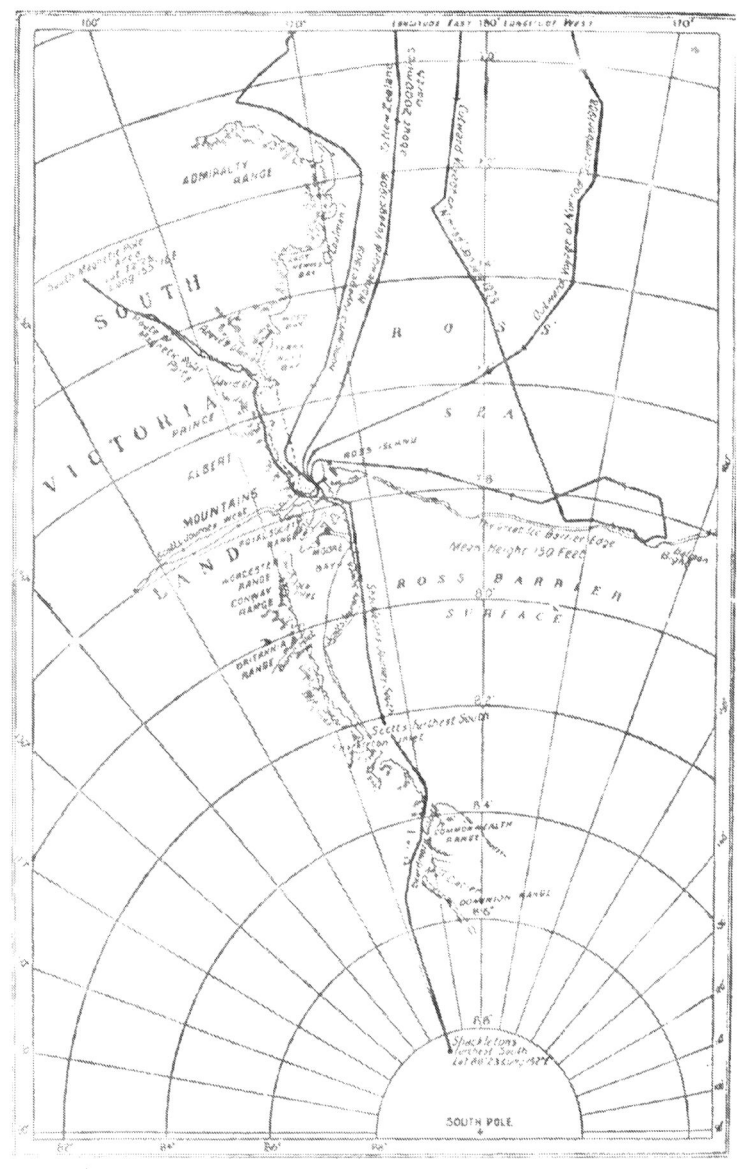

ROUTES OF THE BRITISH ANTARCTIC EXPEDITION, 1907, SHOWING
SHACKLETON'S FARTHEST SOUTH.

CHAPTER III

SHACKLETON ATTAINS. 1908–1909

" Yes, they're wanting me, they're haunting me, the awful lonely places ;
They're whining and they're whimpering, as if each had a soul ;
They're calling from the wilderness, the vast and god-like spaces,
The stark and sullen solitudes that sentinel the Pole "
R W SERVICE.

WHEN the *Nimrod* passed out of sight to the north, Shackleton was left beside a half-finished hut, surrounded by scattered heaps of stores buried beneath mounds of frozen snow, and somewhere out of sight a whaleboat which had been landed from the ship and snowed over. With him were fourteen companions who did not yet fully know their leader, and his chivalrous support of the captain's authority afloat had brought him nearer to unpopularity with his own men than he ever was before—or since. The land party was a group of curiously diverse types, their closest resemblance being in their common love of adventure and desire to penetrate the unknown They came from all stations in life and represented all grades of education

Roberts had been a cook in the merchant service, Joyce and Wild sailors in the Royal Navy, Adams a merchant service and Naval Reserve officer, and Mackay a surgeon in the Royal Navy. These with Shackleton himself made six whose career had been on the sea. Day had been a motor engineer in, England ; Marston, a student of the Regent Street Polytechnic, was a certificated art teacher, and Priestley had been studying geology at Bristol University College. James Murray was a self-taught genius in biology, a man of the rugged Scottish peasant class more thoroughly instructed but in some ways not unlike Thomas Edwards the naturalist, whose life had beguiled one of Shackleton's earlier voyages. If we count Mackay, who had studied medicine at Edinburgh University, over again, there were six University men Of these Brocklehurst, Armytage, and Marshall were Cambridge men, David was an Oxford man by education though Professor of Geology in Sydney N.S.W. and Mawson was a graduate and lecturer of Adelaide University.

As the months went on they all grew to know each other, and they formed a devoted attachment to their leader, whose judgment and foresight in choosing them were well justified. They found before long that the quality which made Shackleton in their opinion too tolerant of the ultra-cautiousness of his second in command afloat, moved him to give each of them a free hand in the pursuit of his special branch of study or allotted duty, and that he never asked them to do harder or more dangerous or disagreeable work than he was prepared to do himself.

By the end of February they had secured nearly all the stores, dumped here and there on the coast round the hut at Cape Royds, at the nearest points to where they had been landed on the ice. This was a difficult task, for during the blizzards the scattered cases had been drifted under snow which partly thawed and then froze, so that each case had to be broken out with ice-axes, the job resembling on a large scale the picking out of whole almonds from almond rock by means of a knife.

The hut, which had no windows, was divided into eight cubicles, four on each side with a narrow open space between. One of these was the leader's cabin, each of the others had spaces for two bunks, in the fitting-up and decoration of which the occupants were left to follow their fancy. It was well on in March before complete order had been reached and the regular routine of life established. The acetylene gas-plant had been got to work, and never before had a polar habitation been so splendidly illuminated. The narrow floor-space was economized by the device of hoisting the long table used for meals up to the roof so as to allow room for work on sledges or harness, or any of the innumerable efforts at construction or repair that were constantly in hand.

THE HUT AT CAPE ROYDS.

Face p. 126.

Before the light went, more than a hundred penguins were killed and stacked up to freeze as fresh food for the winter, and the pony stables and dog-kennels were made as comfortable for their occupants as ingenuity could suggest.

The geographical position of the winter quarters should be clearly understood by the reader in order to allow the events of the various journeys to be followed. Ross Island may be pictured as a triangle each side of which measures over 40 miles. One side runs north and south, Cape Bird being its northern extremity and Cape Armitage its southern. This forms the eastern side of M'Murdo Sound, which is about 40 miles wide and is bounded on the west by the mountainous coast of Victoria Land. Nearly half-way between Cape Bird and Cape Armitage a little flat peninsula tipped by Cape Royds projects into the sound under the steep slopes which rise to the summit of Mt. Erebus. Here Shackleton had established his winter quarters almost exactly in latitude 77° 30' S. and in longitude 166° E. About 10 miles south of Cape Royds a group of islands, named the Dellbridge Islands, lies off the coast, the islands in order southwards being Inaccessible, Tent, Small Razorback, and Large Razorback. These all lie in a bay formed on the south by the projection from the land of a great tongue of ice 5 miles long, tapering from about a mile wide at the base to a fine point: this extraordinary structure, which must be afloat for the greater part of its length, is known as Glacier Tongue. It is 14 miles from Cape Royds, and for 9 miles farther south Ross Island runs in a long narrow peninsula terminated by Cape Armitage. Half a mile north of Cape Armitage on the west side is Hut Point, with the hut of the National Antarctic Expedition at the *Discovery's* winter quarters; and on the east side of the peninsula, about a mile north of Cape Armitage, is Pram Point. The whole west coast of Ross Island, from Cape Bird to Cape Armitage, is so precipitous or so covered with glaciers, ending in ice cliffs, that it is impossible for a sledge-party to travel along it, and very dangerous for an unloaded explorer to make the journey on foot. When M'Murdo Sound is frozen over the sea-ice forms a good thoroughfare along the coast to the south; at other times there is no passage except by sea. The surface of the Ice Barrier is usually found nearly in 78° S., and is attached to the land about Pram Point. The whole southern side of Ross Island from Pram Point to Cape Crozier is wedged firmly into the Barrier, and the north-eastern side from Cape Crozier to Cape Bird is open to the waves of the Ross Sea for the greater part of the year. The extinct volcano, Mt. Terror, rises to a height of 10,750 feet in the eastern corner of Ross Island, and the summit of the active volcano, Mt. Erebus, rises just behind Cape Royds to the height of practically 13,000 feet, sloping steeply on all sides.

Shackleton was intensely anxious to send out autumn sledging parties to lay out depots of provisions on the Barrier to assist the great journey to the South Pole, on which he had set his heart as the central ideal of his life; but the *Nimrod* had hardly gone before the ice on the Sound, the presence of which had compelled him to make his base so far north, broke up and was blown out to sea, thus breaking the road to the south. His sanguine nature rushed to the belief that early spring sledging would enable him to lay out depots enough to help him to the Pole, for nothing broke his happy optimism, yet the truth is that the date of the opening of M'Murdo Sound in 1908 had practically destroyed his chance of success. Had the ice gone out a month earlier he would have landed at Hut Point, two days' journey nearer the Pole, with both autumn and spring available for depot laying; had its break-up been delayed by a few weeks he would have had at least one large depot laid out in the autumn, and the extra food it contained might have turned partial into complete success. In prosperity, Shackleton was somewhat hasty and impetuous, in times of difficulty an amazing patience welled up from the depths of his nature, but only the call of heavy stress could set it free.

He could see only one important piece of exploration possible before the sun disappeared, and the credit of making it he left to Professor David, contenting himself with the less exciting, but no less important task of getting the winter quarters into order. The expedition was to attempt the ascent of Mt. Erebus, a first-class mountaineering feat, the summit being 1000 feet higher above sea-level, from which the climbers had to start, than Mont Blanc is above Chamonix, and, moreover, it is an active volcano. The summit party consisted of David, Mawson, and Mackay, and a supporting party of Adams, Marshall, and Brocklehurst was to accompany them as far as Adams thought advisable. The main party carried provisions for ten days, the supporting party for six. They set out on 5th March, and on the 10th they returned triumphantly, all six having got to the summit in four days and returned in less than one. A severe blizzard had been encountered near the summit, and Brocklehurst had his feet badly frostbitten; but otherwise the exploit had been a brilliant success. The summit of the active cone was found to be 13,370 feet above the sea.

The winter routine was now established. The geologists pursued their studies on the cliffs and amongst the boulders in the neighbourhood of the hut, and here the biologists also found plenty to do in the small freshwater lakes teeming with microscopic forms of life. Even when the water was frozen solid it was possible to make collections by digging shafts through the ice to the lake-bed. The meteorological observations kept Adams busy, and Mawson was able to carry out physical observations on the ice and on magnetism and auroras. Much attention had to be given to exercising and training the ponies and dogs, the number of the latter increasing by several litters of puppies; but, unfortunately, the ponies did not thrive, and several died during the winter from their incurable habit of eating sand.

Inside the hut there was always one member of the party, in rotation, on duty during the day as messman, looking after the housework and responsible for keeping things fairly tidy; another, also in rotation, acted as night-watchman, keeping awake to tend the fire and rouse his companions in the morning. The cook was hard at it all day, for there was no restriction on the food supply, and Antarctic appetites fully justified their gargantuan reputation.

Among the lighter occupations was the preparation of a book, *Aurora Australis*, set up and printed in the hut, with lithograph and process illustrations, also produced on the premises, printed on a hand-press and bound in venesta boards from the packing cases. A hundred copies were printed, but none for sale, and the work is already a rarity for bibliophiles, both on account of the beauty of its typography and because no other printed book has ever been produced on the poleward side of latitude 70°.

A drawback to the general happiness of the wintering party came in the form of Brocklehurst's illness. His frost-bites did not yield readily to treatment, and when it was found necessary to amputate one of his toes, Shackleton gave up his cabin to him and for two months shared with Armytage the cubicle which was the invalid's usual retreat As the long winter night went on, the party became a band of brothers, and the Boss, as Shackleton came to be called, gradually acquired an ascendancy which owed nothing to authority, but was founded on friendship and respect Priestley says of him:

"In the long winter months, when the scientists toiled in darkness and cold at their routine tasks outside, the help and company of our leader might always be relied upon. He was equally at home exercising ponies, digging trenches for the examination of lake or sea-ice, collecting geological specimens, taking the place of an ailing biologist at the dredging line, assisting at a trial run of the motor-car, or breaking in a team of dogs. In the evenings he would retire to his cabin and busy himself writing up his diary, preparing plans for the spring and summer journeys, or writing or reading the poetry which he loved and which he could recite from memory for hours on end. Any jest or argument of unusual interest was sure to draw "the Boss" from his cabin; and when every one else had retired to bed, the night-watchman was never surprised when Shackleton joined him for a half-hour's chat or to smoke a cigarette in the small hours before himself turning in. He was a sociable man and liked company, and was always the life and soul of any group in which he happened to be."

There was no terror of great darkness that winter, nor any depression of spirits in the crowded hut. All the men were well and jolly, always busy and full of hope

From midwinter day, 21st June, onwards Shackleton's

whole life was concentrated on the forthcoming southern journey. In the darkness, dimly lit by the aurora, the call of the solitudes of the icy continent sounded in his ears, whispering to him in the starlit calms, howling for him in the blizzards, always assuring him that he was the man destined from birth to the fame of the discovery of the Pole. He recognized clearly from his experience on the National Antarctic Expedition and from reading that the key of success was transport. The old terror of cold so severe as to extinguish life, and of ice-walls or gulfs so prodigious as to be impassable, had long been laid. So far as geographical knowledge went, it was reasonable to expect that either the Barrier surface, some 150 feet above sea-level, led straight to the Pole, or that if the great mountain chain of Victoria Land swung eastward across the path, some glacier would be found leading to a plateau with as smooth a surface as the Barrier, though some thousands of feet above the sea. In either case success depended on being able to carry enough food and fuel to last a party for the journey of 750 geographical miles to the Pole and 750 back, a total of 1500 geographical or 1730 statute miles.

As a ship must leave the vicinity of M'Murdo Sound not much later than the end of February if she is to get away at all, and as the conditions of travelling over the ice are too severe before the end of October, it follows that four months is the extreme limit of time that can be calculated on for the double journey; and if every day in that period were fit to travel on, the conditions governing success are that food for 120 days can be carried, and that an average speed of $12\frac{1}{2}$ geographical or $14\frac{1}{2}$ statute miles per day can be kept up. As regards food nothing can be left to chance, for the interior of Antarctica is absolutely barren of life; but, on the other hand, there is the advantage as compared with an arid desert or a sea voyage, that no water need be taken, for snow or ice is always present and can be turned into water as long as fuel is available. Economy in transport can only be secured by a system of depots laid out in advance to help the outward journey, or dropped by the way to relieve weights and provide for the return. This is a dangerous method, as the existence of the travelling party must depend on picking up the depot on a featureless waste before the food in hand is exhausted. As regards time, something may be gained if the weather chances to be fine by making a start a few days earlier than it is reasonable to anticipate ; something may be gained on the return journey, if the food supply permits, by overstaying the date for the prudent departure of the ship in the hope that circumstances may allow her to wait longer, or that she has left behind at the base supplies to carry the party through another winter. Should bad weather cause a day's delay at any time, the loss must be made up by longer marches on the good days ; should serious illness or accident befall any member on the outward journey, success would be barely conceivable ; if on the homeward journey, when depots must be picked up at given dates, disaster would be certain. At every point the chances are against success, for the risks are manifold, and any one of them might in a moment end all. A slip on a critical ice-slope, the break-down of a cooker, the loss of a sledge, a slight defect in the sledge-harness or in the alpine rope when a traveller falls into a crevasse, the death of the ponies, a week's blizzard, any of these might make the whole effort vain.

Such were the thoughts that filled Shackleton's mind during the dark months, such were the bases of the calculations he made, the plans he worked out ; and at every report of the illness or death of a pony, all had to be recomputed A weaker man would have wasted his strength and whittled down his chances by giving way to worry, but Shackleton had sublime confidence in himself and his arrangements. He took every possible precaution, the food was carefully chosen, carefully packed, the camp routine worked out to the last detail, and, conscious that he was doing all a man could do, he remained not only tranquil but happy. Priestley says, " Shackleton's faith in his own star never wavered. His optimism and enthusiasm infected his whole following, and went far to make men able to carry on to the very limits of their strength."

On 12th August, ten days before the return of the sun would mark the beginning of the season which it sounds ironic to call the Antarctic Spring, Shackleton set out with David and Armytage on the first preliminary sledging trip, hauling their own sledge, as since only four ponies had survived the winter, it was considered an unreasonable risk to work them before the main journey. They set out on the sea-ice, camped for the first night near Glacier Tongue, and for the next night a few miles from Hut Point.

In the dim twilight of the noontide hours on the 14th, Shackleton led his companions over the old familiar places of the *Discovery* time, to Crater Hill where he used to have his daily walks with Wilson, to the level stretch where he had tried to sail the rum-cart with Barne, to the places where he had helped Hodgson to drag his dredge. Everything was unaltered ; the old hut stood big and bare, the stove gone, and the contents in confusion as they had been left, mixed with snow that had filtered in.

Next day they ascended to the Barrier surface, which stood only 8 feet above the sea-ice, and marched south for 12 miles, camping that night in a temperature below −56° F., cold so extreme that the paraffin used for heating the cooker was of the consistency of cream In this extreme cold, sleep was impossible, and as the weather turned threatening, a hasty retreat was made next day to Hut Point. Here they were kept prisoners by a blizzard for several days, and they passed the time putting the hut in order and building with the cases of provisions an inner hut, occupying 20 feet by 10 of floor space, within which later parties could camp in comfort. On 22nd August they made a fine march of 23 miles to Cape Royds in the one day in spite of adverse weather.

From this time on a party was sent out weekly to Hut Point with stores for the southern journey. This afforded the men excellent training in sledging and camping, and it taught them self-reliance, as the party was not always under an experienced leader. They had often a rough time, as the temperature was very low and blizzards were frequent. The motor-car had been brought into action and did good work in hauling sledges on the sea-ice when free from snow Once with three men it did as much work in one day as six men could have done in two or three days. But the car was useless in snow, and no attempt was made to get it on to the Barrier.

134   THE LIFE OF SIR ERNEST SHACKLETON

The main depot journey began on 22nd September, when Shackleton, with Adams, Marshall, Joyce, Wild, and Marston, set out after saying good-bye to David, Mawson, and Mackay, who were starting in a few days for the Magnetic Pole. The motor towed their sledges at 6 miles per hour as far as Inaccessible Island ; beyond that they hauled, three men to a sledge. On 6th October they reached 79° 36′ S. and made Depot A there, 100 miles south of Hut Point. All that was left in it was a gallon of paraffin and 167 lbs of pony maize Shackleton did not consider it safe to leave there anything that was vital to the expedition, as he was not confident of picking up the depot should bad weather set in The six got back to Cape Royds after covering 320 statute miles in 15½ days of actual marching, an average of 20 miles per day, which augured well for the attack on the Pole.

The summer work had been planned for the two great journeys to the geographical and the magnetic poles. The design of a journey to King Edward Land had to be abandoned as half the ponies had died, and to have divided the four survivors between two expeditions would have secured the failure of both. A third geological party was arranged to the western mountains in connection with depot laying, to assist the return of the Magnetic Pole party.

Shackleton had studied the ponies with the greatest care, and settled the maximum load which each could drag on a sledge. This allowed a maximum of ninety-two days' provisions for the four men who were required to manage the four ponies, so that if the Pole was to be reached on full rations, the daily journey must average 19 statute miles in actual advance. The original plan had contemplated six ponies and six men ; to have taken six men with the four ponies would have limited the radius of action, so with a sore heart the Boss had to decide which two of the appointed six were the least fit and so must remain behind. He felt it a grievous thing to disappoint the ambition of two dear friends, but it had to be On the southern journey of the National Antarctic Expedition the average advance had only been 6½ statute miles per day, estimating distances in the same way without allowing for relaying or deviation from a

CAMP ON THE BARRIER, WITH PONIES.

Face p. 135.

straight course. Thus before he could contemplate a successful journey to the Pole and back, Shackleton had to count on doing three times as well as on his former journey. The prospect did not daunt him. He trusted in his ponies, he trusted in his three chosen companions, Adams, Marshall, and Wild, most of all he trusted in himself. Every possible contingency which his vivid imagination could conceive was provided for.

He arranged that in January 1909 Joyce should make a depot near Minna Bluff (about 60 miles south of Hut Point), with food enough to enable the Southern Party to get back to winter quarters, and that Joyce should remain at the depot to meet them until 10th February, when, if they had not arrived, he was to go back to the *Nimrod*. Murray was left in charge at Cape Royds, and he had instructions to keep a good look out for signals from Hut Point or Glacier Tongue until 10th March, which was judged to be the latest time that it would be possible for the ship to remain. If the ship had to leave before the Southern Party returned, a relief party of three was to be left for another winter at Cape Royds to look for records on a southern journey in the following year.

On 29th October the four men bound for the Pole left the little hut at Cape Royds with the four ponies, two of which were soon found to have gone lame, and with a supporting party of six men who hauled their own sledges. After reaching Hut Point there were delays and some going to and fro before a real start for the south was made on 3rd November. Once on the Barrier a course was laid to the east-south-east in the hope of avoiding the dangerous crevasses which broke the surface of the ice towards the land on the west. Extraordinary care had to be taken, as very often the snow covering a crevasse was strong enough to let a man cross safely, but not to stand the weight of a pony. They had two tents, two men sleeping in each, and the partners changing at frequent intervals so that each could learn to know each of the others equally well. From 12 to 15 miles was the usual march, but on 5th November the two parties were confined to their tents by a blizzard. Next day the supporting party started on their return to the base.

By 13th November Shackleton and his three comrades had got out of the region of crevasses and could relax the extreme tension of the look out ; the worst trouble in going forward now became the soft snow which filled the long hollows between the gentle ridges of the Barrier surface. Shackleton had a severe attack of inflammation of the eyes, which explorers gaily term snow blindness, but it soon yielded to treatment. On the 14th they sighted Depot A, which had been laid out nearly six weeks before, Wild picking it up on camping when sweeping the horizon with his glasses. The upturned sledge and fluttering flag were clearly visible ; and the party slept well, with the justifiable satisfaction of having come straight to this speck on the vast white expanse 60 miles from the nearest land. Shackleton compared it to picking up a buoy in the middle of the North Sea. The weather was fine, the surface fairly good ; for a week their daily march had been 15 miles or more.

On 19th November they found from observations of the sun that they were in 80° 32′ S , a position which was not reached from the *Discovery* until 16th December. This was cheering. The calmness of the weather was extraordinary, although clouds would appear mysteriously and speed across the sky overhead without a breath of wind at the surface. " It is," wrote Shackleton, " as though we were truly at the world's end, were bursting in on the birthplace of the clouds and the nesting home of the four winds , and one has a feeling that we mortals are being watched with a jealous eye by the forces of Nature." The western mountains were now beginning to come into view on the horizon on the right and also straight ahead, so that the route had to be kept still more east of south in order to keep well clear of the crevassed region. On 21st November one of the ponies, Chinaman, which had been weakening for some time, was found unfit to travel farther, and had to be killed. The meat was kept to save the preserved food, and so prolong the time for which the expedition could keep the field   Depot B was constructed, about 100 miles south of Depot A, to make room on the three sledges that were taken on, the material left being 80 lb. of pony maize, a 27-lb. tin of biscuit, some sugar, and a tin of paraffin, stores which would allow the four men food enough to take them back to Depot A on the return journey. This meant that the lives of the party depended absolutely on their being able to pick up this speck in the wilderness a couple of months later. Now their eyes were gladdened by the sight of land rising into mountain peaks which had never been seen before. Shackleton's diary says:

" The land consists of great snow-clad heights rising beyond Mt. Longstaff, and also far inland to the north of Mt. Markham. These heights we did not see in our journey south on the last expedition, for we were too close to the land, or rather, foothills, but now at the great distance we are out, they can be seen plainly."

The Barrier surface was as flat as a billiard table, the ponies pulled well, the daily marches lengthened to 18 miles, and on 26th November they passed the farthest south reached in 1902, which it had taken the *Discovery* party, starting at the same date, until 29th December to reach. That night there was a carouse in the tents to celebrate the breaking of the record. A four-ounce bottle of curaçao sent by a friend at home, and the now rare luxury of a smoke set free the springs of conversation, and for the time the four felt that their goal was all but won. Some glimmering of what the leader spoke of may be caught from a soliloquy in *The Heart of the Antarctic*:

" . . . It was with feelings of keen curiosity, not unmingled with awe, that we watched the new mountains rise from the great unknown that lay ahead of us. Mighty peaks they were, the eternal snows at their bases and their rough-hewn forms rising high towards the sky. No man of us could tell what we would discover in our march south, what wonders might not be revealed to us, and our imaginations would take wings until a stumble in the snow, the sharp pangs of hunger, or the dull ache of physical weariness brought back our attention to the needs of the immediate present. As the days wore on and mountain after mountain came into view, grimly majestic, the consciousness of our insignificance seemed to grow upon us."

On 28th November Depot C was made in latitude 82° 40′ S., and here a second pony, Grisi, had to be shot. A week's provisions and oil, as well as a quantity of the horse-flesh, were left here, enough to carry the party back to Depot B, and a sledge was left as a mark. The onward march was made with two loaded sledges, each weighing 600 lb., including supplies for nine weeks, and two men helped each pony by hauling on the sledge. The Barrier surface was now ridged into long, low undulations, which were only discovered by the disappearance and subsequent reappearance on the northern horizon of the snow pillars built at each resting-place to serve as guides for the return. On 1st December, in 83° 16′ S. latitude, a third pony, Quan, Shackleton's special favourite, was found quite worn out and had to be shot, leaving only Socks to help with hauling

Pulling the sledges was now heavy work for the men, as the weather continued calm and the sun shone hotly day and night, high above the northern horizon at noon and not very much lower above the southern horizon at midnight. The air temperature rose nearly to the melting point of ice, and the four men, stripped to their shirts, suffered from the heat, finding comfort only in chewing raw frozen horse-flesh, which cooled them though it did little to assuage the savage hunger which was intensified by the unending toil. They sighed or cursed, according to habit, at the thought of what they could do if only there was food enough to eat as much as their natures craved for; but there was no rebellious thought, and Shackleton was greatly strengthened in his resolution by the good will with which his fellows faced harder work and poorer fare in order to increase the chance of reaching the Pole. The mountains were trending more and more to the eastward, and the time had come to strike due south in search of a way across their ramparts

An isolated summit, some 3000 feet in height, appeared close ahead, and naming it Mt. Hope, Shackleton shaped a course which brought him to its base on 2nd December. Next day he left the sledges and pony at the camp, and with his companions clambered across a stretch of ridged and crevassed ice and round an enormous chasm, 80 feet wide and 300 feet deep,

Camp on the Beardmore Glacier: Site of the Lower Glacier Depot.

to the granite rocks of Mt. Hope. The coloured goggles he wore made it difficult to see, and with a characteristic impulse he took them off, saw clearly, but paid for it with a sharp attack of "snow blindness." This he counted as nothing because of the vision of the road to his promised land which burst upon him on scaling the precarious weathered granite rocks of his Pisgah. Before him rose great bare mountains with prodigious cliffs falling sheer for thousands of feet to a stupendous glacier which descended between them from a high snowfield far to the south and lay like a road to the Pole, smooth and straight and gently sloping. Distance made nothing of details of structure which were soon to assume gigantic proportions. Where the ice-river met the plain of the Barrier it gave rise to tremendous disturbances, séracs, pressure ridges, crevasses, dying off ultimately into the long undulations which they had detected on their way from the north. To this grand feature Shackleton gave the name of the Beardmore Glacier, in honour of the friend whose guarantee had made the discovery possible.

Next day, by almost superhuman labours, they pioneered a way over a gap 2000 feet in height, whence they descended to the glacier surface with their pony and two sledges. Here the Lower Glacier Depot was made on 5th December at a camp close under a gigantic granite pillar, from the weathered surface of which the travellers, accustomed so long to be themselves the highest figures in the landscape, were fearful that falling stones might bombard their tent.

As they pursued their way up the Highway to the South their difficulties increased. The pony had great trouble on the patches of smooth ice, and Wild was unfailing in attendance on the poor beast, leading it as far as possible along the snow-slopes. The morning of 7th December had been full of anxiety on account of the shadowless half-light due to cloudy weather which conceals all irregularities of surface. Later in the day the light improved and the party, now about 1800 feet above sea-level, proceeded cautiously, Shackleton, Adams, and Marshall pulling one sledge in front, Wild leading the pony with the other sledge behind them. A sudden cry of " Help ! " stopped the advance, and, turning round, the three men saw the pony-sledge sticking up out of a crevasse, with Wild holding on to it ; but Socks, the pony, was gone. It had broken through the snow-bridge, and, but for the fact that its weight in the sudden shock of the fall had snapped the swingle-tree of the sledge, Wild and the stores would have followed it into the abyss.

Had the sledge gone it is doubtful if the party, with only half its equipment, could have got back to Hut Point. The loss of the pony, which was to have been shot that night, did not appear so serious at the moment ; but as events developed it seems that the want of the meat from this pony was one of the factors which made the attainment of the Pole impossible. At the time Shackleton only felt that the accident made his task more difficult, and he still had reserves of mental strength enough to believe that harder work on his part might still lead to complete success. Next day and the next they pushed on with improving surfaces for 12 miles each day along the glacier, rising by night to 2500 feet. The utmost vigilance was still necessary, for one of the party fell through a snow bridge into a crevasse and escaped death only because his sledging harness and the sledge itself held under the shock. Had they failed, he would have gone to the bottom of a crack which was estimated to be 1000 feet deep. Before they were done with the Beardmore Glacier, every one had repeated this experience. There had been no scamped work on the equipment or in the testing of it ; otherwise not one of the party would have returned alive.

Good progress continued until 11th December, when the party had got up to 3300 feet ; but they were still 340 miles from the Pole, and straining every nerve to push forward on reduced rations. In the midday rest Shackleton and a companion climbing the lower slopes of the mountains bordering the valley in which their glacier highway ran, noticed that old moraines lay in terraces far above the actual surface of the ice, and gathered geological specimens from the rocks in place—small ones, to be sure, for every ounce to be carried was a consideration, yet enough to set the geologists to work on their return. The two left at the camp utilized the rocks that lay about them to grind up a quantity of pony maize by crushing it between stones so that it could be used to eke out their provisions

The slope of the glacier now began to steepen, and progress had to be counted rather in feet of ascent than in miles of advance, while the task of hoisting the sledge over the rugged and slippery ice-falls was becoming heart-breaking. The deplorable system of relaying had to be adopted for the first time on this journey, and in three days only 15 miles were gained though 45 were travelled ; but the height above sea-level had risen to 5600 feet. Each day Shackleton hoped that the end of the glacier was at hand and that an easy run over the Plateau would follow. But every day new mountains continued to emerge on the west and east, and Marshall kept taking angles to them and mapping the country as they proceeded. This work was extremely difficult and laborious, and was carried out throughout the journey with consummate skill, yielding a map of quite unusual accuracy for such a march. Every day they fell many times ; but though any fall might have had fatal consequences, no serious damage was done. On the 17th the remarkable discovery was made that certain dark bands which ran along the face of a huge sandstone cliff were seams of coal though apparently of poor quality. There was no time to spare for a close inspection, and as the Plateau was at last in sight, or so they thought, it was time to prepare for a final spurt of speed as soon as the level could be reached. A depot was accordingly made in 85° S , about 6000 feet above sea-level. All the heavy clothes were left here and provisions for four days, that being considered sufficient to bring them back to the depot near the lower end of the glacier which they had left ten days earlier ; they hoped when going downhill to move twice as fast as in climbing, or if not, to eat less in proportion. One of the two sledges was left here and one of the tents, all four crowding into one, the floor space of which was completely covered by their sleeping bags.

The slope of the glacier became less, but a week had yet to pass with an average rate of advance of 10 miles per day before the crevasses which beset the upper snowfield were left behind and the Plateau fairly entered on. Some days the

142 THE LIFE OF SIR ERNEST SHACKLETON

frozen surface rang hollow under their feet as if they were walking on the glass roof of a station. The weather continued good, but the short rations kept all the men in a state of unceasing hunger, with the thought of food never out of their conscious minds by day and dominant in their dreams at night They had been scrimping themselves by an extra effort to have some semblance of a feast on Christmas Day. The day came and they were at last on the edge of the Plateau at a height of 9500 feet, almost at the 86th parallel of latitude, still 250 geographical miles from the Pole, and Shackleton was still confident that it could be reached with enough reserve of food and strength to make a safe return to the sea possible The evening meal came at last. Shackleton says in his book·

"We had a splendid dinner. First came hoosh, consisting of pony ration boiled up with pemmican and some of our emergency oxo and biscuit. Then in the cocoa water I boiled our little plum-pudding, which a friend of Wild's had given him. This, with a drop of medical brandy, was a luxury which Lucullus himself might have envied; then came cocoa, and lastly cigars and a spoonful of *crême de menthe* sent us by a friend in Scotland. We are full to-night, and this is the last time we will be for many a long day. After dinner we discussed the situation, and we have decided to still further reduce our food. We have now nearly 500 miles, geographical, to do if we are to get to the Pole and back to the spot where we are at the present moment. We have one month's food, but only three weeks' biscuit, so we are going to make each week's food last ten days. We will have one biscuit in the morning, three at midday, and two at night It is the only thing to do. To-morrow we will throw away everything, except the most absolute necessities."

Next day they lost sight of the mountains, and their horizon was limited by a circle of snow on the bare, upward-sloping Plateau. On each of three days they made a march of 14 miles and reached 86° 31' S. at a height of 10,000 feet. Now Shackleton hoped that with fine weather he could reach the Pole by 12th January and make a rush back to catch the *Nimrod* on 28th February. The great altitude was having its effect, however, and this there was no means of counteracting. One after another began to suffer from frightful headaches ; the wind blew stronger, and their worn clothing and tent no longer kept it out ; deficient food had lowered their vitality, so that their temperatures were all 4° subnormal. Only their unconquerable will kept them going on, farther and farther from adequate food and safety. One day a blizzard reduced their progress to 4 miles ; but on New Year's Day, 1909, they passed latitude 87°, and had beaten all previous records towards the Poles, north or south. There was now nothing in life but the one struggle to get on in the teeth of difficulties, and difficulties were piling up from the storms outside and the growing weakness within. All were on the verge of frost-bite, and the greatest torture was experienced through the freezing of the moisture from their breath or the water running from inflamed eyes on their faces and beards, often caking into one solid mass with their clothes, and in the effort to reduce weight they had not left themselves with even a pair of scissors to cut the matted hair about their mouths. On 2nd January Shackleton wrote :

"God knows we are doing all we can, but the outlook is serious if this surface continues and the plateau gets higher, for we are not travelling fast enough to make our food spin out and get back to our depot in time. I cannot think of failure yet. I must look at the matter sensibly and consider the lives of those who are with me. I feel that if we go on too far it will be impossible to get back over this surface, and then all the results will be lost to the world. We can now definitely locate the South Pole on the highest plateau in the world, and our geological work and meteorology will be of the greatest use to science ; but all this is not the Pole. . . . I must think over the situation carefully to-morrow, for time is going on and food is going also."

Two days later ·

"The end is in sight. We can only go for three more days at the most, for we are weakening rapidly. . . . We started at 7.40 a.m , leaving a depot on this great wide plateau, a risk that only this case justified, and one that my comrades agreed to, as they have to every one so far, with the same cheerfulness and regardlessness of self that have been the means of our getting as far as we have done so far. Pathetically small looked the bamboo, one of the tent poles with a bit of bag sewn on as a flag, to mark our stock of provisions which has to take us back to our depot, 150 miles north. We lost sight of it in half an hour, and are now trusting to our footprints in the snow to guide us back to each bamboo until we pick up the depot again."

On 6th January, Marshall got an observation of the sun which placed them in 88° 7′ S., or 113 geographical miles from the Pole, and the diary says, "To-morrow we *must* turn"; but the bitter wind against which they had been fighting for days rose to the highest fury of a blizzard, and all the morrow and the day after the four men lay shivering in their tent, unable to face the blast and unable to keep their minds from dwelling on the risks of their footprints being obliterated, of the depot being snowed under and the flag carried away . . . "it is a serious risk that we have taken, but we had to play the game to the utmost, and Providence will look after us."

Even then, with nothing to read, with almost nothing to eat, and with nothing to do save occasionally to nurse the frozen foot of a comrade back to life, the spirit of Shackleton flamed high; he pictured the triumph of their return, even without the culmination of their hopes, and he strove in spite of his own bitter disappointment to hearten the others.

On 9th January 1909 the weather cleared; the four men left the camp and sledge at 4 a m. and half-running, half-walking, they came by 9 a.m. to a point which was estimated as in 88° 23′ S and 162° E., 97 geographical or 113 statute miles from the South Pole. They displayed the Union Jack presented by Queen Alexandra, their personal sledge flags had been left behind in the Glacier depot to save the few ounces of weight, and Marshall took a photograph of the other three standing beside the flag. A brass tube was buried in the snow with a record and a sheet of Antarctic stamps; the Plateau was formally annexed to the British Empire, and taking the flag with them these men, the remotest from their kind in all the world, turned their backs on their unreached goal and started their

FARTHEST SOUTH, 1909. ADAMS, WILD, SHACKLETON, AND THE QUEEN'S FLAG.

Face p. 144.

race of 700 miles, with Death on his pale horse, the blizzard, following close.

To their relief they found their footprints clearer than when first produced. The snow compressed beneath their weight had resisted the wind which had swept off the loose surface, leaving the tracks marked out in little blocks several inches high. The fierce wind continued to howl behind them day after day, and a sail hoisted on the sledge was a great help in hauling On the 10th and 11th the distance made averaged 18 miles ; the depot left on the Plateau with such misgiving was safely picked up, and Death was beaten on the first lap. As the surface began to slope downward the pace improved, until from the 14th to the 17th they were doing more than 20 miles a day, though suffering from continual hunger, and kept from sleeping at night by the pain of their lacerated heels. The sledge-meter by which they had measured the distance travelled got broken, and distances had to be estimated and checked by observations of the sun.

The mountains came into sight on the 16th, and as the snowfield merged into the glacier the speed increased to 26 miles on the 18th and 29 miles on the 19th. This was the best day's march ever made, the sledge being rushed down ice-falls and over crevasses, and sadly strained on the way.

The depot laid on the Beardmore Glacier in latitude 85° S. on 17th December was picked up on 20th January and the four days' provisions there secured, as well as the second tent and sledge, so that the conditions of camping were improved a little, though the labour of tent-pitching was increased. Death was beaten on the second lap. The day had been full of trying experiences, the slippery, clear blue ice, swept clean of snow, led to frequent falls, and Shackleton was very badly shaken, so disabled indeed, that for a whole day he could not haul, and it was as much as he could do to keep up with the sledge. Delay was impossible, for after supper on the 25th there was only one short meal left, which was eaten on the morning of the 26th, and the party went on fasting for the rest of that day and all the next until the Lower Glacier Depot left on 5th December was reached in the evening. Death was beaten this time by a margin of hours only. These were the hardest and most trying days of the whole expedition, and the men, dauntless even in starvation, came to the very limit of endurance. They were marching or rather scrambling amongst ice-falls and crevasses for nearly twenty hours with nothing but a cup of tea or cocoa to support them, constantly breaking through snow bridges, and only saved from destruction by their sledge harness. Shackleton wrote in his diary: "In fact only an all-merciful Providence has guided our steps to to-night's safety at our depot. I cannot describe adequately the mental and physical strain of the last forty-eight hours." It is a fact vouched for by all the four men that even then there was never a cross word from one of them nor the semblance of a quarrel on the whole journey.

With a spare sledge-meter from the depot and six days' food to carry them over the 50 miles to Depot C on the Barrier, they got off the Glacier to their intense relief and set their faces homeward on the last but longest stretch of their desperate race. The weather was not too favourable, but that was not the most disquieting circumstance. Some of the party had developed dysentery, it was supposed from eating the horse-flesh from the depot.

Wild, who had been the first to be stricken with dysentery, was unable to eat the horse-flesh, and suffered horribly from hunger. At breakfast-time a biscuit was served out to each, which could be eaten at the time or kept till later in the day. On 31st January Wild finished his at once, and as he was starting on the march he found Shackleton's hand slipping a biscuit into his pocket "What's that, Boss?" he asked, and the answer was, "Your need is greater than mine." He resisted; but Shackleton was irresistible and fought in silence with his hunger, for he knew his friend was more hardly put to it than himself. The other two men never knew of the incident. No one could say that Shackleton was acting the part of Sir Philip Sidney for his own glory, for until now the facts were written only in Wild's private diary. There he says, "S. privately forced upon me his one breakfast biscuit, and would have given me another to-night had I allowed him. I do not suppose that any one else in the world can thoroughly realize how much generosity and sympathy was shown by this ; *I do, and BY GOD I shall never forget it."* He never did, as the record of their great friendship abundantly proves.

The snow pillars marking the outward route were standing and were spotted one by one, ensuring a straight march on the Grisi depot (Depot C) which had been left on 28th November. This was reached on 2nd February, the birthday of Shackleton's boy Ray, and in honour of the day two extra lumps of sugar were served out. A start was made next day with a new sledge, in deep soft snow, and progress was slow. The meat from the pony Grisi was perhaps poisoned by some ptomaine, it was surmised by the toxin of fatigue, for the animal was in an exhausted condition when shot. Every one was down with dysentery on the second day out. The ninety-one days for which the party had been provisioned on leaving Hut Point were past, and there was no alternative to the horse-meat as staple food for weeks to come. Delay was more dangerous than ever, for the *Nimrod's* skipper, whoever he was, must by this time be looking anxiously at the ice-conditions and might not be able to await the coming of the Southern Party. 4th February was calm and sunny, but no one could stir from the tent, and the day was passed in unspeakable misery. Death had crept up terribly close this time. Next day the men were a little better, and marched 8 miles, then followed days with 10 and 12 miles of northing made; some of the party still felt very ill, all were extremely weak and hungry. The wind strengthened from the south and for four days the sail helped them along, enabling them to make 16 to 20 miles daily on half a pannikin of half-stewed horse-flesh and five biscuits per man each day.

On the evening of 12th February the flag on Chinaman depot (Depot B), which was left on 21st November, was sighted. It was reached next morning, and the liver of the pony was "splendid"; a solid red mass found by Shackleton in the snow puzzled them for a time, then it was recognized as frozen blood and proved a welcome addition to the diet. For days past all the thoughts of every one had been of food; but eating was a torment, for the worst trouble that had now come upon them was that of blistered and burst lips due to the cold wind and their enfeebled state. On the 15th, Shackleton's birthday, his comrades prepared a present that was a complete surprise. It was a cigarette made of the last shreds of their pipe tobacco rolled in coarse paper; but the kind thought of his friends lent it a flavour more appreciated than the dainty Tabards of his luxurious days.

On the 17th a terrible southerly blizzard, which would have kept them prisoners in their tent at any other time, was hailed as a friend, and with the sail up they made 19 miles that day, venturing to increase their rations somewhat on the strength of the good march, and often hardly able to keep ahead of the sledge as the roaring wind drove it on. Still the dominating obsession of all was food. "We all have tragic dreams of getting food to eat," writes Shackleton, "but rarely have the satisfaction of dreaming that we are actually eating. Last night I did taste bread and butter. We look at each other as we eat our scanty meals and feel a distinct grievance if one man manages to make his hoosh last longer than the rest of us."

On 18th February Mt. Discovery was sighted, on the 19th the familiar outline of Erebus with its flag of steam, the winning-post of their race, came in view, and on the 20th they reached Depot A, which they had left on 15th November, ninety-seven days since. They enjoyed the pot of jam originally intended for Christmas, which had been left here to save weight, and they had now plenty of tobacco; but all their hope of getting back to safety depended on the food in the depot which Joyce was to have laid out to the east of Minna Bluff. He had orders not to stay there after 10th February, and by this time he was probably back on the ship.

The finding of Joyce's depot was a precarious adventure, the worst if almost the last obstacle of their dreadful race. There were no tracks to lead to it, and it was impossible to be certain that it was there at all During the next two days a blizzard raged, into which only starvation could drive the most valorous of travellers; but the direction was favourable. Death would overtake them if they halted, and they made 21 miles a day in it. Even so, by the second evening they had finished their food. Then unexpectedly they struck the track of a sledge which led them to a deserted camping-place where empty tins with unfamiliar labels proved that the ship must have returned to Cape Royds before Joyce's party left to lay the depot, and that the depot had been laid. A diligent hunt amongst the rubbish in the snow found nothing eatable but three little pieces of chocolate and a scrap of dog-biscuit. Lots were drawn as to which man should have each fragment, and the Boss had the ill-luck to fare worst, as the half-gnawed bit of biscuit fell to his share. Next morning early, 23rd February, Wild caught sight of the longed-for depot flag, far away on the right and only raised into view above the horizon by the accident of a mirage which faded away just as Marshall had taken its bearing. But for this miraculous help, so it appeared to them, the depot would certainly have been missed ; Death would have caught them, and the tragedy of three years later to Scott's party would certainly have been anticipated.

The line of march was changed; at 4 p.m. the depot was reached with abundant food and even a superfluity of luxuries. Never were starving and exhausted men more richly succoured. Warned by experiences on the *Discovery*, Shackleton preached moderation to his companions and practised it himself. The letters from the ship, which had reached M'Murdo Sound on 5th January under the command of F. P. Evans, formerly of the *Koonya*, gave rise to some anxiety as to how long she would stay, and Shackleton had now a new inducement for fast travel ; but next morning Marshall was found to be very ill, and a blizzard was blowing in which it would have been madness for him to move The day was spent in sleeping bags, with ample food, a welcome rest to tired bodies though fraught with anxiety for doubtful minds. On 26th February they did a 24-mile march on which Marshall suffered greatly though he stuck to it doggedly. Next evening after another march he was so ill that Shackleton left him in the tent with all the food and Adams to look after him, while he and Wild with one day's rations and a light sledge made a forced march to Hut Point. To their horror they found open sea 4 miles south of Cape Armitage, and had to travel

150   THE LIFE OF SIR ERNEST SHACKLETON

7 miles round before a landing could be made on the east side of the peninsula far north of Pram Point.

Late on the night of the 28th they staggered up to the *Discovery* hut, a hundred and seventeen days after they had left it True they were back on the very day Shackleton had decided on, as the latest date for his return, when at the farthest camp on the great Plateau ; but there was no time for congratulating himself on that score   There was no one at the hut, only a letter reporting the return of the Magnetic Pole party with their task accomplished and the safety of all the other members of the expedition, which was good news ; but going on to say that the *Nimrod* would be lying at Glacier Tongue *until 26th February*, and this was very alarming. It was now 28th February, and if the ship had gone the fate of Marshall and Adams was sealed and their own also. Here at the very end of the fateful race Death had pulled up abreast of them, and the issue had never been more doubtful. There was food and fuel in the hut, and before long Shackleton had a meal ready ; but they had left all the sleeping gear with the sledge when they made the last effort to reach land, and there was no covering in the hut but some old roofing felt ; wrapped in this they sat up all night, too cold to sleep. They made an attempt to set the old magnetic hut on fire in the hope of attracting attention on the ship—if she had not gone—but they could not get it to burn. They climbed to Vince's cross to hoist a flag on it, but their numbed fingers could not tie a knot. On the march Shackleton had often talked of what he would do if the ship had gone, and he had even made a plan for sailing to New Zealand in the whaleboat. This effort was not necessary. When the darkness passed they tried the signals again and succeeded, and in a short time the age-long hours of watching ended and the ship appeared. They were on board the *Nimrod* at 11 a.m. on 1st March, received like men returned from the dead, as they had been given up for lost by the captain, who had only been persuaded by the unanimous appeal of the Cape Royds men to delay his departure to the last possible moment.

Worn as he was with his four months of hardship and the two last sleepless nights, Shackleton himself led the relief party

BACK FROM THE FARTHEST SOUTH.
WILD, SHACKLETON, MARSHALL, AND ADAMS ON THE *NIMROD*, MARCH 1909.

to take in Marshall and Adams, setting out for the south again, three hours after he had reached the ship, with Mackay, Mawson, and M'Gillan, and in less than twenty-four hours they arrived at the camp and found Marshall better. When they got back to the ice-edge late on 3rd March, the ship was not to be seen. Another difficult landing had to be made to reach Hut Point, where a carbide flare brought the ship back about midnight. That happened in this way. Mackintosh had been pacing the deck of the *Nimrod* with a companion when he suddenly said that he felt certain that Shackleton had returned to the hut. His friend scoffed at the idea that any one would march in such a blizzard, and jokingly said that Mackintosh should go up to the crow's nest and wait until he saw a signal. Mackintosh went aloft and saw the flare as soon as he reached the masthead.

Shackleton and Adams got on board at one o'clock in the morning and found the captain intensely anxious to get north out of the reach of the young ice. They were too utterly exhausted to return, and Adams had to be cared for, as he had fallen into the sea when getting on board the boat from a slippery ice-ledge. Marshall and the others were sent for in all haste, and the hut was closed and left, though time did not admit of going back with tools to make good the damage of years of stormy weather. There was no neglect shown by Shackleton's party.

The ship steamed round Cape Armitage to beyond Pram Point in the afternoon, and a party landed to secure the sledge with geological specimens from the Beardmore Glacier which Shackleton had left there four days before and felt it right to run some risk to recover. Then the *Nimrod* turned and made all speed to the north, passing Cape Royds, where the hut had been left with a full year's provision for fifteen men; the key hung outside the locked door, and a letter inside placing the whole at the disposal of any future expedition. Two years later Priestley, then on Scott's last expedition, revisited this hut and found that a meal of bread and tongue, which had been left unfinished on the table, was still in perfect condition.

152 THE LIFE OF SIR ERNEST SHACKLETON

There were still some days fit for exploration, and Shackleton determined to utilize the time to the best advantage by trying to follow the coast of the continent westward beyond Cape North. The *Nimrod* passed Cape Adare on 6th March, and by the 8th she had got off Cape North in longitude 166° 14' E. and in latitude 69° 47' S. A stretch of coast was seen running west for about 45 miles with every appearance of being the northern end of the great Plateau. The eyes of Mawson were turned towards it to some purpose, and he marked it out as the scene of future Australian exploration The young ice was forming rapidly, and Shackleton was just in time when he decided at midnight on the 9th to escape from the danger of another wintering, yet there were some anxious hours in getting back to open sea. Then a short and pleasant passage brought them to New Zealand, affording time for complete recovery from all the fatigue and starvation. One member of the Southern Party put on twenty-eight pounds of weight in fourteen days

Shackleton's mind, freed from the tension of constant thought on the insistent details of life at the very limits of human endurance, had time to accustom himself to the larger bearings of his expedition. He had justified to himself and to all men his family motto, *Fortitudine Vincimus*, and he recognized that he had not only done his best, but done supremely well. As of old, the albatrosses swooped about his ship, ever watchful, and he could meet their scrutiny with pride and a clear conscience no matter what sea-hero's soul might animate them, for he had proved himself to be in the direct succession of Cook, Biscoe, Balleny, and Ross, the discoverers of the Farthest South.

The arm-chair geographer, who is apt to take cold-blooded views, saw two great results from Shackleton's southern journey. One, that he had advanced so far, the other, that he had not reached the Pole ; that sentimental but powerful magnet still remained to attract new expeditions, for which the conditions of success had been shown to be simple and the task comparatively easy The greatness of Shackleton's advance is apparent from this little table, showing how each successive expedition outdid its immediate predecessor in poleward progress :

Explorer.	Date.	Highest S. Latitude.	Years later than preceding.	Miles farther than preceding.	Geog. Miles distant from Pole.
Cook	1774	71° 10'	—	—	1130
Weddell	1823	74° 15'	49	185	945
Ross	1842	78° 9'	19	234	711
Borchgrevink	1900	78° 50'	58	41	670
Scott	1902	82° 17'	2	207	463
Shackleton	1909	88° 23'	7	366	97
Amundsen Scott	1912	90° 0'	3	97	0

Thus Shackleton has the greatest advance to his credit. He also, building on experience with the National Antarctic Expedition, pioneered the way to the Plateau in high southern latitudes. He clearly grasped the fact that transport was the key to the whole problem, and he took the most original and effective steps to overcome all difficulties. To his organizing power and to his generosity of character must also be placed a large share of the credit of the two great exploits of his expedition in which he did not take part personally, the ascent of Mount Erebus and the journey to the Magnetic Pole, in which Sir T. W. Edgeworth David and Sir Douglas Mawson proved themselves to be worthy to rank with the foremost polar explorers of all time.

Perhaps the greatest triumph of all was to bring back the expedition without the loss of a single life or permanent injury to the health of any member ; and this must not be put down to chance. Shackleton certainly was amazingly fortunate in escaping disaster, not once or twice, but almost daily. He and his followers seem to have led a charmed life ; but, looking closely, we may see that in almost every case escape and safety followed on a quick decision, as, for instance, if the ship should push on or turn ; if the party should march or camp during a blizzard ; if, when sledging, they should rush a crevasse or pull up on the brink; if they should leave four or six days' provisions in a depot, and the like. Shackleton's mind combined many diverse and sometimes contradictory qualities; his keen imagination and instantaneous Irish intelligence gave his mental processes the quickness of intuition, but the solid Yorkshire qualities of shrewdness and caution always rose to the surface in emergencies, stimulated out of their Saxon deliberation by the urge of Celtic fervour. So it seems that Shackleton's quickest decisions were fully reasoned out from solid data, but so rapidly that the process eluded himself in a way similar to that by which a practised wireless operator reads the signals he receives directly as words, without consciously spelling them out letter by letter. Making full allowance, however, for the working of a finely-balanced and quickly-acting brain, which goes far to explain good luck in the ordinary affairs of life, we must recognize that again and again there was no assignable reason for a crisis leading to success instead of failure, and Shackleton's belief in providential guidance was perfectly sincere.

The engagements with newspapers at home for exclusive news on which large payments to the funds were contingent, required some careful management to avoid premature disclosure, and the explorer had perforce to become man of business again The place appointed for sending the first dispatches was the most southerly telegraph office of the Dominion, the quaint little settlement of Oban, situated on Half Moon Bay in Stewart Island, and separated from the main part of New Zealand by a wide and stormy channel where only one small steamer plied once a week.

The ship reached the island on 22nd March, some hours too late for the appointed time of cabling, so she anchored on the south of the island and the whole party landed to have a picnic tea on a lonely beach under the marvellous growth of tree-fern and creepers which gives to this outpost of the Empire the colour and the luxuriance of the tropics. Back from the blank and blinding whiteness of the snow and the blackness of cliff and shadow which had alone diversified the scene for the last fifteen months, all felt the gracious presence of the gentler powers of nature as no visitor to that land had ever done before.

They watched with delight the wekas or Maori hens which walked out of the undergrowth to share their meal, as unconscious of danger as penguins, and at night they were lulled to sleep by the cry of the little owls persistently calling " More pork, more pork ! "

Next morning the *Nimrod* steamed round to Half Moon Bay. At 10.30 Shackleton landed in his own boat, which immediately put off from the shore, and he walked alone under the almost arboreal growth of the southern gorse to the little post-office. In ordinary circumstances telegrams are telephoned across the channel to The Bluff ; but for this message a telegraph instrument had been specially installed and a skilled operator had been waiting since 10th March. Shackleton had a long account of the expedition's doings written out on telegraph forms on board, and after sending a brief code message to announce his return, the great news was put on the wires. Next morning four columns appeared in the London papers without a single mistake in word or figure, and there was no leakage. Probably only a few personal friends were thinking of Shackleton on 23rd March On 24th March he was the talk of the whole civilized world, and when the *Nimrod* entered Lyttelton Harbour on the afternoon of the 25th, she met a blizzard of congratulations. How the boy who counted his letters when the *Hoghton Tower* reached port would have gloried in the masses of letters and telegrams—and the man on the *Nimrod* was still a boy at heart ! No wonder that his narrative of the whole expedition concluded :

" The loved ones at home were well, the world was pleased with our work, and it seemed as though nothing but happiness could ever enter life again."

CHAPTER IV

POPULARITY. 1909-1910

Nor will I say I have not dreamed (how well!)
    Of going   .   .   .   forth,
As making new hearts beat and bosoms swell,
    To Pope or Kaiser, East, West, South, or North,
Bound for the calmly satisfied great State
    Or glad aspiring little burgh . . .
        where learned age should greet
My face, and youth . . . he learning at my feet . . .
    With love about, and praise, till life should end,
And then not go to heaven, but linger here,
    Here on my earth, earth's every man my friend—
The thought grew frightful, 'twas so wildly dear!
    But a voice changed it."
                                ROBERT BROWNING.

WHATEVER dreams of recognition and praise might have cheered Shackleton during the year of struggle to get his expedition together, and his year of toil and endurance in the farthest South, they fell short of the reality. No traveller, possibly no man, ever woke up to find himself so suddenly and so universally famous. No man stood the shock better either; the essential modesty and generosity of his nature enabled him to keep his head, and much as he enjoyed the sunshine of popularity it changed in no way his love for his own people and his devotion to old friends. He entered with the keenest zest into the year of popularity which lay before him. He was young enough to feel to the full the gratification of honours conferred, while friends of youth were about him to share the happy glow.

The first great public function he attended with most of his followers was a Thanksgiving Service in the fine cathedral which helps to give to Christchurch its peculiarly English aspect, and here by special request he heard once more the hymns which had long had for him a mystical meaning as associated with his own career—" Fight the good fight with all thy might," and " Lead, kindly Light "—songs of work and faith.

The King truly expressed the feelings of the Empire in this telegram :

"I congratulate you and your comrades most warmly on the splendid result accomplished by your expedition, and in having succeeded in hoisting the Union Jack presented you by the Queen within 100 miles of the South Pole, and the Union Jack on the South Magnetic Pole. I gladly assent to the new range of mountains in the far south bearing the name of Queen Alexandra.
" EDWARD R. AND I."

Amongst the tributes telegraphed from the home papers he valued most highly the generous words of his old leader, Captain Scott (whose great nature revealed no grudge at the change of plan forced upon Shackleton by the conditions of the ice round King Edward Land), before he could have received any personal explanation. In fact, when the two met the subject was not referred to, each waiting for the other to raise it. One of the finest incidents in the history of exploration, the continued good relations between two men who might so easily have become envious rivals, passed unnoticed, because no one beyond a very small circle knew the circumstances.

There was much work to be done in New Zealand arranging for the homeward voyage of the *Nimrod*, which, under the command of her former first officer, John King Davis, had a fine programme of hunting for dubious islands marked on the chart, but never seen since their reported discovery. Arrangements had also to be made for the development of photographs and kinema films, for the preparation of the book which was to help so greatly to pay the debts of the expedition, and for a multitude of other details. And as he worked Shackleton was always parrying the assaults of the reporters with smiles and jokes instead of the copy they strove for. Twice he travelled

158 THE LIFE OF SIR ERNEST SHACKLETON

to Wellington, first to thank the Prime Minister for the fine help and generous reception New Zealand had given to the expedition, then on the eve of sailing to lecture in the Town Hall to 3000 people, and put £300 into his pocket—only to pass the whole sum over to local objects with which he sympathized.

Crossing to Australia at the end of April he was acclaimed by crowds, received by governments and universities at Sydney, Melbourne, Adelaide, and Perth, gave lectures in each capital to thousands of enthusiastic hearers, and with the reckless generosity of his nature handed over hundreds of pounds, tendered to him as fees, to the local hospitals. He swept through Australia on a roar of applause, and sailed in the same liner, the *India*, in which he had travelled out. On the voyage home he worked at his book, dictating much of it to a young New Zealander, Mr. Edward Saunders, who accompanied him to England as a literary assistant. He talked quietly to sympathetic fellow-passengers, and took pleasure in romping with the children; and at the ports of call he used to keep the reporters waiting until he had finished the game he had entered on with his little friends. At Port Said he transferred himself with the mails to the *Isis*, and was rushed across to Brindisi by his old friend and comrade of *Discovery* days, Captain A. B. Armitage, who was then in command of the mail boat. In Italy he had a foretaste of the strenuous days before him. His publisher, Mr. Heinemann, was waiting for him and accompanied him home, planning the details of the great book which was to be published within the year. The Royal Italian Geographical Society sent a deputation to inform him that he had been awarded its great gold medal; and, rushing through France, he leaped ashore on Dover pier at 3 o'clock on the afternoon of 12th June and found his wife waiting for him where she had seen him off twenty months before.

For forty-eight hours the reunited couple disappeared from the world. Mrs. Shackleton had continued to live in Edinburgh with the children and her sister, and there in March she had been struggling with growing anxiety as the days went on, until on the forenoon of the 23rd her sister had persuaded her to go to the baths for a swim, her favourite recreation. So she had no idea that even then an early special edition of the evening paper was shouting the great news in Princes Street. As she got home telegrams were beginning to arrive from friends full of congratulations, amongst them a little cable from New Zealand, "Absolutely fit. . . . Home June"; and now it was June and home. They motored out together on the Sunday to the old house at Tidebrook and wandered in the woods, recalling happy days and talking of the expedition, though Shackleton always made light of his share in it. He made no complaint even of the hard luck in having to turn back when the goal was in sight, summing it up in the remark, "Anyway, darling, I thought you would like a living donkey better than a dead lion."

The London season had decreed, all the same, that a living lion he was, and insisted that he should play the part, and the people of England and of the world were of the same mind. In the next twelve months Shackleton had to fulfil more engagements and meet more people than in all his life before. We can only hope to give a faint idea of the blaze of light in which he and his wife lived during those enchanted months, or of the crash of song in which his deeds were celebrated. The first month may be looked at in some detail as a sample of what the hero of an hour has to go through.

Long before 5 o'clock on the afternoon of Monday, 14th June 1909, the great gates of Charing Cross station had to be closed to keep out the crowd from the forecourt, and the arrival platform was already filled with personal friends. There was Dr. Shackleton, most of his family, his small grandson Ray in a sailor-suit with *Nimrod* on his cap, the smaller granddaughter Cecily, and many relatives. There was Captain Scott, foremost to greet his old comrade; Major Darwin, President of the Royal Geographical Society, with many members of Council and officials, including several polar veterans, and a host of others, privileged to offer the earliest greetings to Lieutenant Shackleton, as the public still delighted to call him. No one who was present is ever likely to forget the roar of cheering from the crowd which filled the Strand and Trafalgar Square as the open carriage, with Shackleton, his wife and children, made its way slowly along the streets where no attempt had been made to keep a passage open, for the police had failed to foresee this burst of enthusiasm. The newspapers put the crowd at 10,000, and telegraph boys had delivered 400 telegrams at the door by the time the party arrived.

Next day there was a lunch given by the Royal Societies Club, with Lord Halsbury in the chair, when speeches of welcome and approval were made by Lord Halsbury, Sir Clements Markham, the promoter of the *Discovery* expedition, and Sir Arthur Conan Doyle. Shackleton's reply was a masterpiece of spontaneous oratory, rising to the height of a great occasion. The same evening Mr. Heinemann gave a dinner in the Savoy Hotel where the leading literary men of London were present, and here the explorer displayed in his speech the fascination which the finer forms of literature always exercised upon him. Sir W. Robertson Nicoll, one of the guests, said of this speech, " He is a born speaker and a born leader of men. With perfect self-possession, with easy command of himself and his audience, with an ample choice of fitting language, and with a rare modesty, he summed up the lessons of the expedition." Another literary journalist of great experience in judging the age and qualities of public men, who was present, thought that Shackleton was not more than thirty years of age—an estimate five years short of the mark, but a tribute to the boyish spirit which always rose triumphant after the hardest experiences. Next night there was a superb Society dinner and reception in Park Lane, where Tetrazzini sang; the next a dinner of a City Company, and the week ended with a night of Bohemian jollity at the Savage Club, where Captain Scott presided and Shackleton solemnly signed his name on the wall behind the chair, to be glassed in later like the autographs of Nansen and other great travellers which kept it company.

The following week was filled with a crowded succession of lunches, garden-parties, dinners and receptions, for the most part at the mansions of great ladies, who were determined to have the lion of the season on exhibition; but a few at the homes of old friends to whom he was no lion, but a man with whose efforts they had sympathized and in whose triumph they rejoiced. Several honours and distinctions had been conferred by geographical and other societies abroad ; but the first of the first-class distinctions at home to fall to Shackleton's lot was his appointment as a Younger Brother of the Trinity House, with the special approval of the Master, the Prince of Wales. This honour is peculiarly valued by merchant-service officers, and it appealed in particular to one who revered the ancient authority which watched over lighthouses and the navigation of our coasts.

On Monday, 28th June, the Royal Geographical Society had arranged for a special meeting in the Albert Hall at which the first detailed description of the British Antarctic Expedition should be given by its leader and the photographs and kinematograph films exhibited. Before the meeting the Geographical Club, a body more ancient than the Society itself, entertained Shackleton and all of his comrades who were in England to a dinner at which the Prince of Wales (our present King) was present, together with all the polar travellers then available. The meeting was a brilliant success About eight thousand people, presided over by Major Leonard Darwin, filled the vast hall; but Shackleton's great voice filled it too, and as he spoke, easily and without notes, modulating his tones to suit his theme, he held the immense audience hanging on his words. As he had once on a time held the *Hoghton Tower* close to the wind with his hand on the wheel, so now he brought the people shaking on the brink of tears, then with a scarcely perceptible touch let them fall away in laughter. And this was now no unconscious power, but an art in which he knew and enjoyed his mastery. In that supreme hour he remained calm and deliberate, unexcited and free from agitation or affectation. It was observed that he hardly ever used the pronoun " I," but always " we," and no leader ever acknowledged more fully the part played by his comrades. For the first time those who had never travelled near the far South saw the movements of the penguins, as they pursued their quaint ceremonious lives, with their own eyes on the screen ; they saw the ponies harnessed and the sledges moving off towards the Pole, the motor-car struggling on the sea-ice, and

162   THE LIFE OF SIR ERNEST SHACKLETON

the *Nimrod* charging through the pack, and were able to realize how tremendous was the revolution made in the art of polar travel and description by the British Antarctic Expedition, 1907. The lecture over, the Prince of Wales rose, and with a few words of hearty appreciation of the lecture and the lecturer concluded, "as a brother sailor I am proud to hand him this medal " ; presenting the special gold medal which the Society had had struck for the occasion, showing Shackleton's portrait on the obverse. His Royal Highness then handed silver replicas of the medal to those other members of the expedition who were able to be present, namely, Adams, Armytage, Brocklehurst, Day, Joyce, Mackintosh, Marshall, Marston, Murray, Priestley, Roberts, and Wild, the presentation being made in alphabetical order.

Two days later the lecture was repeated in the Queen's Hall to the general public, with Lord Strathcona in the chair. Another week of dinners from lords and ladies famous in Society held some more interesting functions sandwiched between the lion-hunts. The Mayor of Lewisham, the "glad aspiring borough" which had grown to include Sydenham, welcomed him in the name of his townsfolk; the American Society honoured him at their Independence Day banquet ; and he was called upon to open the Exhibition of Travel and Sport at Olympia.

On 12th July the Shackletons were commanded to Buckingham Palace, where King Edward and Queen Alexandra received them very graciously, heard the story of the great journey with interest and attention, and the King bestowed upon Shackleton the third class of the Royal Victorian Order, following the precedent in Captain Scott's case after the return of the *Discovery*

For the three weeks between 7th and 29th July Shackleton's secretary typed out a list of thirty separate engagements, including the opening of flower shows, prize givings at several schools, of which his own Dulwich College was one ; dinners at City Companies and at great houses and famous clubs. One Saturday night he visited the Crystal Palace to see fireworks designed to represent himself at the Farthest South, after which he caught the midnight train to Leeds to address a great Sunday meeting of four thousand in the Town Hall in aid of the National Lifeboat Fund. The week before he had snatched a week-end in the country to get on with his book, at which he was working steadily for hours each day in the intervals of his fixed appointments Towards the end of the month he had a meeting entirely after his own heart at the Browning Settlement in Walworth, where he was presented with the badge of the Settlement, bearing the words from his favourite poem, " Prospice "—*Sudden the worst turns the best to the brave*; and where he hailed the working men as brothers; for, as he told them, he had been a worker ever since he shovelled coal at Iquique on the deck of his first ship. Amongst these people he found himself at his best, because he was in touch both with reality and poetry as he never was at the grand Society functions July ended with the reward common to fame and notoriety of a realistic effigy in Madame Tussaud's.

The months of August and September were full of hard work on *The Heart of the Antarctic*, the writing of which was finished in an amazingly short time; they were full also of pleasant visits and happy surprises. For the third time Shackleton was in the Solent for the beginning of Cowes week, not this time outward-bound on an expedition awaiting Royal inspection, but with his wife as Sir Donald Currie's guests on·board the fine new ship of his old line the *Armadale Castle*. Here Sir Donald had extended his historic hospitality to a great many important and interesting people. Amongst others of his new friends Shackleton met on board Sir Henry Lucy (" Toby, M.P."), who was greatly interested in the expedition and was so much moved on hearing that a debt of nearly £20,000 lay on the leader's shoulders that he sent a signed statement of the case to a London daily paper on 5th August. This raised a wave of public sympathy, and a general desire that something should be done. It did more than this, because on reading the article the Prime Minister, Mr. Asquith, sent for Shackleton on the very day it appeared and assured him that steps would be taken to help him.

A fortnight later Mr. Asquith wrote to Shackleton :

" With reference to interviews I had with you on the subject of the liabilities incurred in connexion with the British Antarctic Expedition commanded by you, I am pleased to be able to inform you that the Government have decided to recommend Parliament to make a grant of £20,000 to meet a portion of the expenditure. The Government have been induced to take this course as they are much impressed, both by the great value of the discoveries made in the course of your voyage and by the efficient and economical manner in which the whole of the enterprise was conducted, as is shown by the fortunate return of your entire party, and by the comparatively small total outlay incurred."

The news was received with a chorus of approval, both in Parliament and in the Press. The whole cost of the expedition had been under £45,000, less, indeed, than had been paid for the ship alone of the National Antarctic Expedition, 1901, and Shackleton was now able to relieve his guarantors at once from their obligations and pay off the debt at the bank. In his characteristically exuberant way he saw fortune as well as fame in his grasp, and declared that from the book and the lectures which had been arranged, he would make £50,000 for himself : an absurd overestimate, of course, but one which set the scale for his lavish expenditure and unmeasured generosity to relatives and the charities he delighted to surprise with his benefactions. If only he had had a strain of thrift in his character his future might, indeed, have been assured in comfort ; but the fact must be faced that, always grasping at fortune, Shackleton never secured himself against pecuniary embarrassment, though in this happy year of success the wolf of anxiety was far from his door.

Early in September the news arrived that Dr. F. A. Cook claimed to have reached the North Pole, and was being fêted in Copenhagen ; and later came the announcement of Admiral Peary's arrival at the North Pole. Each denied that the other had got there, and a great controversy began in which the general public on both sides of the Atlantic took sides.

King Edward commanded Shackleton to deliver his lecture

at Balmoral, and there he enjoyed hearty Highland hospitality for a delightful week-end and was in no way awed, as this letter to his wife shows :

" I am up here safely, and it is very comfortable. The King is very jolly, and took me down last night to arrange about the lecture and went into all details himself. There are some interesting people here, and one dear old naval man, 89. Lord Rosebery and Sir Allen Young and Slatin Pasha are here also. . . . I sat next but one to the King last night. . . . He wore Highland costume. . . . We had a very good dinner and did not turn in until 11.30 p.m. The King enjoys a joke very much. He asked me a lot about Cook and Peary and Scott. He seems to know everything that is going on."

On the suggestion of the King himself, we understand, Shackleton was elected to the Marlborough Club, where to the end of his life he was a well-known and popular member.

The *Nimrod* had returned, and it was suggested that she might be thrown open to the public in the Thames. Shackleton sprang at the idea, but instead of leaving her in a dock below London Bridge, he decided to have her moored at the Temple Pier and an exhibition of Antarctic specimens and equipment in a hall on the other side of the road. Before the amazed riverside men could point out the impossibility of taking such a ship under the bridges, Shackleton had the masts out of her, the hull— so small in the Southern Ocean—looming gigantic as she was towed through London Bridge, the Cannon Street railway bridge, the Southwark Bridge, and the two bridges at Blackfriars. Then her masts were set up again and her crow's nest looked down at Cleopatra's Needle. Thousands of people visited the ship during the month she lay there, and £2000 was secured from their shillings for the London hospitals before the *Nimrod* set out on a tour round the coast in aid of charity. Anybody could have done a thing like this ; but only Shackleton did.

In October he made a tour of the Scandinavian capitals with his wife, and had his first taste of foreign Court life. They travelled in sleeping cars, and, as guests, they occupied the finest suites of rooms in all hotels. The lecture to the Geographical Society in Copenhagen was attended by the King and Queen of

Denmark, Queen Alexandra, the Dowager Empress of Russia, and many other royalties of Denmark, Russia, and Greece, all of whom chatted amiably with Mr. and Mrs. Shackleton. The audience of nearly two thousand followed the lecture in English and laughed heartily at the jokes. Next day, after royal and diplomatic receptions, Shackleton repeated the lecture for the benefit of the poor of the city. Commander Hovgaard, the great Danish Arctic explorer, took charge of the arrangements and formed a close friendship with his young colleague of the South. They enjoyed every moment, and he carried away the order of Commander of the Dannebrog. He was presented by the Geographical Society with a gold medal, or rather with the case for it, as the medal which had been prepared for him had been given to Dr Cook a month before and the new one was not yet ready.

So on to Stockholm and then to Christiania, always greeted with popular acclaim on arrival, and hurried to palaces and legations for pleasant meals with kind and hearty kings and queens, or kind and courtly ministers, always waited on and hailed as an equal by the most famous explorers of the country, always listened to with rapt attention by crowded audiences, always leaving with a new gold medal and a fresh illustrious order. In Stockholm, Dr. Sven Hedin and Professor Montelius entertained them, in Christiania Dr Nansen and Captain Roald Amundsen, in Gothenburg Dr Otto Nordenskjöld, all men who had had long experience of polar conditions and knew what a man had come through in pushing into high latitudes. The enthusiasm was perhaps greatest in Christiania. There the students, led by Amundsen, conducted Shackleton in a torch-light procession from his hotel to the lecture hall, and afterwards carried him shoulder-high, with the short barking cheers, nine times repeated, that express so well the fervour of the roused Scandinavian spirit. At one banquet Mrs. Shackleton records:

"I shall never forget the look on Amundsen's face while Ernest was speaking. His keen eyes were fixed on him, and when Ernest quoted R. Service's lines, ' The trails of the world be countless,' a mystic look softened them, the look of a man who saw a vision."

That was the moment she believed that Amundsen resolved to turn the voyage on the *Fram* which he was then preparing from the North to the South Pole. As they were leaving Christiania, a young man, almost a boy, jumped into their carriage and had a long and earnest talk about the South polar regions; this was Tryggve Gran, who afterwards was taken on the expedition which Scott was then avowedly preparing.

The progress swept on to Brussels, where the reception was magnificent, though at that moment the Court was somewhat constrained towards British visitors on account of criticisms in the London Press on Congo administration. King Leopold had his grim joke in offering Shackleton the Congo medal, which he felt obliged to decline; but Prince Albert (the present popular King) presided at the lecture with tactful sympathy, and nowhere was there heartier applause. Captain de Gerlache, who had faced the first Antarctic night in his ill-found little *Belgica*, their host, translated the lecture into French, sentence by sentence, and amongst those most appreciative was the venerable Madame Osterrieth, whose devotion had made the dispatch of the *Belgica* possible in 1897, and who still smiled to hear her affectionate nickname of those days, "Mère Antarctique."

At the end of October *The Heart of the Antarctic* was published in two great volumes, containing a full account of the British Antarctic Expedition, with appendices on the scientific work and the richest array of illustrations that ever graced a book of travel since the photographic era came in. As a feat of rapid writing and expeditious printing, this work was un-rivalled, and it showed no signs of haste in style of writing or form of presentation. Professor David's great journey was described by himself, and both parts of the work revealed the individuality of the authors.

Then began the great series of lecture tours which filled the next twelve months. As designed, it was to include lectures at one hundred and twenty-three different places in Europe and America, involving 20,000 miles of travelling, and the addressing of audiences totalling a quarter of a million. Mr. Gerald Christy, the veteran lecture agent, had never before laid out so large a programme for any public man; but Shackleton tackled it as he had tackled his Southern journey, and carried it through without damaging his health or wearing out his popularity. The first fifty lectures filled November and December. Four were in Scotland, three in Ireland, forty in England The chairmen were often men of high distinction: the splendid old Arctic sledge traveller, Admiral Sir Lewis Beaumont, presided over the opening of the campaign in London on 1st November; other chairmen included a Viceroy, an Archbishop, the Headmasters of the greatest public schools, Dukes, Earls, Viscounts, Judges. But usually it was His Worship, the Mayor, whose introductory and valedictory oratory had to be borne once by the audience, but many times by the lecturer; full of goodwill it always was, but rarely lit by humour, though with such a lecturer as Shackleton this merely enhanced his brilliance.

There were some pleasing incidents by the way At Rugby on 9th November, the headmaster was able to congratulate the lecturer on the honour of knighthood announced that morning in the King's Birthday List. At Edinburgh, ten days later, on receiving the Livingstone Gold Medal from the Royal Scottish Geographical Society, Sir Ernest was able to say that as secretary he had sometimes carried a Livingstone Medal to that meeting, but he had never had to carry one away from it before. At the Irish lectures in Dublin, Cork, and Belfast he was received as an Irishman, and rose to the occasion to the delight of his audiences. At Halifax he was almost snowed up, and had to improvise a sleigh to take him over the blocked roads to reach the hall, giving him a homely feeling, he said, to be in a blizzard again.

The English lectures were interrupted for a few days to allow of a visit to Paris with Lady Shackleton, where Prince Roland Bonaparte presided at the meeting of the doyen of Geographical Societies, and presented the singularly beautiful medal. Here Shackleton took the bold step of reading his lecture in French, and his courage was warmly applauded, coming as it did in the first flush of the Entente Cordiale. Moreover, the French translation of his book, *Au cœur de l'Antarctique*, had been published in Paris the day before. In describing the evening the *Figaro* said. "When the applause ceased at last the explorer, in a clear and well-toned voice, began to read his lecture, which was written in excellent French though pronounced with a pretty strong British accent, which, however, did not prevent the audience from grasping the sense of the words, so sustained was the attention with which they listened." There were receptions by the Municipality of Paris in the Hôtel de Ville, and by the President of the Republic in the Elysée, a dinner from the British Ambassador, and a banquet from the British Chamber of Commerce. Despite his "accent un peu difficile," Shackleton was a popular hero, and when decorated with the Legion of Honour he almost felt himself a Frenchman, remembering his Huguenot ancestors.

One more high distinction at home was paid to the Expedition by the King personally decorating Sir Ernest and his companions of the shore party with the Silver Polar Medal, or an extra clasp for those who had been on the *Discovery*, and the officers of the *Nimrod* who had not wintered, with the bronze medal.

The New Year, 1910, opened with a continental lecturing tour, sixteen lectures being given in twenty-two days in Italy, Austria, Hungary, Germany, and Russia. The tour began in Rome, where Shackleton, who always hated sight-seeing of the ordinary sort, was with difficulty persuaded by his wife and her sister to look into St. Peter's, and conducted to the spot just as the doors were being closed in the evening. This escape from the boredom of being shown round delighted him extremely. He went through the lecture with pleasure, though it was read in French by another, and he was received most cordially and decorated by the King. The Pope, strangely enough, was not moved to inquire into the Antarctic continent, and there was no visit to the Vatican.

The next appointment was in Berlin, where a formal reception was arranged at the arrival station. Changing at Munich, by some mischance, the Shackletons found themselves in the wrong train and reached Berlin before the appointed time, however Shackleton had to return to the station, apologize for his premature arrival, and go through the ceremonies according to plan. There were two lectures to immense and delighted audiences. On the first evening he spoke in English at the Berlin Geographical Society under the genial presidency of Professor Penck, with the Crown Prince and Princess in the forefront of the audience. On the second night he courageously essayed to read in German to the Colonial Society, where Herr Dernburg, the Colonial Minister, presided with great grace and cordiality The delivery of this lecture was by no means so happy for himself or his hearers ; but despite the " disastrous German " of his Dulwich College days, he persevered, as his Quaker ancestor Abraham Shackleton had persevered with Latin, and before long could make himself perfectly understood. The goodwill with which he grappled with the language ensured him immense popularity. There was much feasting at formal parties, lasting four hours at a stretch, with speeches between the leisured courses.

Then on to Vienna, where, on 9th January, the Imperial Geographical Society gave a grand reception and he a fine lecture, made the more memorable by the sharp contrast between the lonely tent on the Plateau where he lay nearest the Pole on that night in 1909, and the brilliant company he faced in 1910 with its galaxy of Archdukes and Archduchesses.

At Budapest the Geographical Society was even more effusive, and the hospitality if possible even more kind. So powerful was it, that Mr. Locsy actually overcame Shackleton's resistance to sightseeing, though he could not conquer the repugnance which showed itself in the lugubrious face in contrast with the malicious triumph in the eye of the too kind host. Among the Hungarians he made many friends, and laid the foundations of future visits and high hopes.

At Hamburg the triumph was repeated before a huge audience, who recalled the lecturer four times, and on 14th January, only a week after his first lecture in Berlin, he was back in the capital for two public lectures. More interesting was a private lecture given in English at Herr Dernburg's house, for the special benefit of the Kaiser, who appeared in the uniform of a British Admiral of the Fleet, and was greatly interested and most genial He had his views on Antarctic exploration, believing that Shackleton would have got on better if he had kept more to the west. With this the explorer, regardless of etiquette, bluntly disagreed, and he was also obliged to answer the well-meant question, " Did you shoot many bears ? " with " There are no bears in the Antarctic, Sir," only to be countered by " Why not ? " It was a very friendly meeting all the same, and resulted in mutual feelings of esteem between two strong men. Of all the German royalties Shackleton liked best Prince Henry of Prussia, whose frank recognition of a brother sailor was irresistible ; and when, after lecturing at Breslau, Dresden, Leipzig, Frankfurt, and Munich, he came to Kiel on 22nd January, he was entertained by the Prince as his guest in the Castle.

The next engagement was with the Imperial Russian Geographical Society in St. Petersburg, and Prince Henry gave a personal letter of introduction to the Tsar, while Princess Henry gave one to the Tsaritsa Thus heralded, the visit to the Russian capital in the midst of its winter splendour went like a fairytale. He was introduced at the lecture by the venerable vice-president, M. Semenoff-Tianshansky, who had been honoured in the Russian way by hyphening the field of his explorations to his name, and entertained also by General de Schokalsky, the leading Russian oceanographer and polar student. Before leaving, he had a long audience of the Tsar, who talked to him for nearly two hours. It seems a paradox to assert that Shackleton's Antarctic journeys were less remarkable than his progress through Europe ; but it is certain that any strong young man at any time in the future may, if he can command a few thousand pounds, revisit the Great Barrier, the Beardmore Glacier, and the vast Plateau leading to the Pole, and find all these exactly as Shackleton saw them, as Scott and Amundsen saw them ; but to no man now living will it ever be possible, however much wealth he may be prepared to lavish in the quest, to find again the proud Court of a Hohenzollern in Berlin, of a Habsburg in Vienna, or of a Romanoff in St. Petersburg : dynasties which have passed "with their triumphs and their glories and the rest," and capitals which are desolate or bankrupt.

Six days after lecturing in St. Petersburg, Shackleton was addressing an audience in Aberdeen, and in the seven weeks following he lectured in forty other towns from St. Andrews on the north to Hastings on the south, from Lowestoft on the east to Plymouth on the west. Wherever he went he found enthusiastic audiences to whom the story of his adventures in the South were new, or at least the pictures were; and to a man so attuned to sympathy as Shackleton, the tale he had told a hundred times was fresh every evening with some slight shade of difference drawn forth by the different hearers. Even so, and notwithstanding that all trouble over the arrangements was taken by his highly-skilled agents, the strain must have been heavy. Yet in between the formal lectures he found time for special addresses for charities, for interviews strongly vindicating the integrity and polar success of Peary, and once to preside at the annual Court of the Seamen's Hospital at Poplar. When, accompanied by Lady Shackleton, he went on board the *Lusitania* on 19th March, he must have hailed the week's rest before him as a boon.

The voyage was delightful, and, of course, Shackleton gave a lecture on board in aid of a nautical charity. The chair was to have been taken by Mr. de Navarro (the husband of Mary Anderson), who was going to America to visit his mother, whose death was made known to him by wireless the day before the lecture. A few hours after hearing the news Shackleton sought out his friend, who writes as follows:

"He appeared in my cabin, the vibrant voice tuned to a low key. I expressed my regret at having to fail him. 'It doesn't matter,' he answered quietly, ' I have come to do something for you and to ask you to do something for me: there will be no chairman to-morrow night. That for you. For me,' he hesitated a moment. ' I want you to come to the lecture.' Amazed at the request, I was silent. He went on, ' I'll put you behind the screen where no one will see you and where you will see and hear all.' Though deeply touched I told him that I preferred to remain in my cabin. ' No,' he interrupted warmly, ' that is exactly what I want you to avoid. I want you to come and do violence to yourself at once, without losing time, and so begin building up the strength necessary to bear what you will have to face when you land.' The persuasive words, the searching eyes, the solicitous insistence, I had not the power to resist. The lecture ! a disciplinary struggle, rousing examples of courage, optimism, endurance in Antarctic solitudes (a seeming personal appeal in each), a sense of self-reliance. And the landing ! If it was not attended by the measure of fortitude he had hoped for, at least in supreme moments the face of a viking with a mother's heart came between me and my grief."

It is good to have this testimony of a man in a high degree sensitive and sympathetic himself, to the vein of tenderness and delicacy which Shackleton's character contained hidden altogether from the notice of ordinary people.

The American lecturing tour began in Washington on 26th March in circumstances parallel to those of lectures in the European capitals. The Shackletons were entertained by Mr. and Mrs. James Bryce at the British Embassy; they were received at White House by President Taft, who later presented to Shackleton the Hubbard Gold Medal of the National Geographic Society at a crowded meeting of 5000 persons who listened with enthusiasm to the explorer's lecture. He visited Miss Wilkes, the daughter of the only great American Antarctic explorer, Admiral Wilkes, who was a contemporary and rival of Sir James Ross. At New York, Admiral Peary presided at the lecture to the American Geographical Society and presented the Cullum Gold Medal, while the stay in the metropolis was glorified by the magnificent hospitality of the New York Four Hundred. At Philadelphia the Geographical Society, under the presidency of Mr. H. G. Bryant, Peary's main supporter, provided another great and hearty audience.

At Boston there was appreciation and a large hall, but it was not full for Tetrazzini was singing, not this time in his honour, but as a rival at the Opera House. The students of Harvard went mad over Shackleton, who never forgot the fervour of the College yell. But then the tide suddenly fell; the smaller towns of New England in which it was arranged that he should lecture made no response. For a night or two he spoke to empty benches, and then discovered that the ground had not been prepared by the lecture agents. There had been almost no advertising; he arrived in one town in the same train as the posters announcing his lecture, which should have been adorning the walls for a week before. This was more than Shackleton could stand, he took firm action after plain speech, and interrupting his American programme he crossed to Canada and spent a few delightful days with his wife as the guests of the Governor-General and Lady Grey.

On the forenoon of the day when he was to lecture at Ottawa he was stopped as he had often been by an important-looking personage with a paper in his hand. "Sir Ernest Shackleton?" inquired the stranger. Shackleton, recognizing the leader of a deputation, acknowledged his identity with all modesty and his most attractive smile. "That's all," said the stranger, thrusting into his hand a writ issued at the instance of his American agents prohibiting him from exhibiting his kinematograph films in Canada! The Canadian Court quashed the injunction and the lecture went well and smoothly, the chair being occupied by Sir Wilfrid Laurier, the eloquent Prime Minister. The Canadian poet Robert Service had long been a favourite of Shackleton, who said :

"Often I have quoted Service, because he is the one man in the world who brings home the glamour and the mystery of the unknown. Reading his poems one can understand how Canada claims men, and how they ever must follow her lure."

The Shackletons returned to the States, and for a fortnight they remained in New York or Philadelphia, meeting people of importance and interest, dining one day with Mr. Choate to meet Lord Kitchener, on another occasion delighting the Press Club with an impromptu speech which Mr. Jerome, the great lawyer, commended highly, saying that it must have cost much care and labour to prepare ! On another occasion the luncheon party consisted entirely of the most successful commercial men in New York, millionaires who did not think their time wasted in making the acquaintance of the young explorer

Shackleton was very much disgruntled, as his new friends

would say, by the breakdown of his plans, though the fault was not his, and at times he was tempted to throw up his Western tour and return home. But he stuck to it all the same One day in New York he walked back to his hotel in pouring rain from a place 4 miles away, and when expostulated with, explained that he couldn't find a taxi, wouldn't go in the subway, and thought the tramcars looked dangerous; in fact, he exemplified unconsciously his preference to rely on himself in an emergency.

The Western tour, on the advice of a new agent, Mr. Lee Keedick, who soon became a valued friend, was cut down to three weeks, for the lecturing season was coming to an end. Shackleton mistrusted the conditions likely to be met with, and was glad when Mr. and Mrs. Craig Lippincott of Philadelphia insisted on Lady Shackleton remaining as their guest. He left her there in all the luxury that wealth and taste enable cultured Americans to provide, while he faced the unknown with his usual determination to see it through. The lectures took place in some of the chief centres of the States of Ohio, Iowa, Indiana, Michigan, Illinois, Nebraska, and Minnesota. In some places, as in Dayton, Ohio, where there was a splendidly-organized reception, and in Chicago, everything went well. In others there was bad management locally and great apathy; in one huge hall, seated for 4500, the audience numbered only 100. In one hall the electrician refused to put the lights up and down, because he said he wanted to listen to the lecture and not to be disturbed. The lanternists were sometimes incompetent and impudent, and the lecturer was thoroughly disgusted, and in the phrase which recent successes had made a great stranger in his vocabulary, he declared that he was "sick of it all." There was at least one gleam of sunshine in the West, in the good cheer at the University of Michigan at Ann Arbor, which Professor W. H. Hobbs refers to in these terms:

"I was deputed by the student committee in charge of his lecture to meet him. I found him at first almost grouchy over the reception he was having. While people in Europe had been turned away from his lectures for lack of space, he had been speaking to small audiences in the States, and these lacked enthusiasm. This was due to the Cook fiasco. To use a slang expression, the American public was 'fed up' with polar exploration He said to me, 'The American people have no sense of humour,' and you know what a keen sense of it he had. But a surprise was in store for him, for here in Ann Arbor a packed house awaited him. I could see the change in his temperament the moment he looked out on them. He first threw on the screen a map of the Antarctic with the 'Farthest South' of earlier explorers, and pointing to that of Cook, he turned to the audience and remarked dryly, 'No relation.' Again, to use American slang, 'the roof rose' with the instant roar of applause. I could see his face light up, and I knew he was mentally modifying his earlier declaration concerning American sense of humour. At a dinner before the lecture, he had sparkled in telling us stories of the winter in camp on M'Murdo Sound at the winter quarters, but he had indicated much concern over the operator of the lantern, evidently having had some unpleasant experiences When Sir Ernest was putting on the wonderful 'movies' of the penguins at the close of his address, he chanced to glance up at the balcony to the lantern, and was thunderstruck to see the roll of precious film on which the success of his remaining lectures depended, instead of being wound up on the reel, piled on the floor in a loose pile already visible above the balustrade. He stopped his talk and shot some language at the operator, which had all the force if not the words of profanity. Had a spark from the lantern got into the pile we should have had a conflagration. I rushed up to the gallery so as to shoo the crowd away as the meeting broke up We found among the students a former expert on lantern work and with his help got the precious film back on the reel without serious damage. Hardly was this done when news came of the death of King Edward, and I learned through the effect on Sir Ernest the meaning of the sovereign in Great Britain to a loyal citizen. He was terribly depressed and felt the loss most keenly."

It was with real pleasure that the Shackletons, after meeting at Chicago, found themselves in Canada Shackleton was once more in the full light of popularity. The lecture at Winnipeg on 21st May was given in his best style to an audience of 1200 enthusiastic citizens. He visited the hospital where his sister

Eleanor had been working as a nurse, and her friends greeted him the more warmly because of his striking resemblance to her. The President of the Canadian Pacific Railway placed his private car, the *Nanoose*, with a clever French cook, at the disposal of the hero of the day and his wife ; and in this they travelled for eight days, being hitched on to the train each day and left in a siding convenient to the place of lecturing each night. So Brandon, Regina, Calgary, Edmonton, and Vancouver were visited in state—everywhere delighted audiences and good houses (at Vancouver the proceeds were £240 for one lecture), and everywhere amusing incidents. At one town, when Shackleton was waiting for his wife at a shop door, a young man said, " You're not doing anything. Hold my horse for a minute, will you ? " " Certainly," was the reply. The minute turned out to be fifteen, and on being thanked with " Next time I meet you I hope you'll come for a drink, what's your name ? " the explorer said, " Thanks. Shackleton." The young man beamed and said, " Then I'd like to shake," and shook hands cordially. There was nothing of the stuck-up Englishman, so hateful to the western Canadian, about Ernest Shackleton.

At Winnipeg, on the way back, he gave another lecture in the largest hall, where the audience included 1200 soldiers in uniform and 2000 civilians ; the proceeds were given to the Children's Hospital where his sister worked. Tired but glowing with satisfaction, the little party travelled east, too tired, indeed, to lengthen their journey by a day to see Niagara. Sir Thomas Shaughnessy entertained them in princely fashion at Montreal, and in his private car, *Killarney*, they took the train to Quebec, where on 10th June they sailed for home in the Allan liner *Virginian*.

The Canadian trip had suggested the idea of undertaking exploration in the Arctic lands to the north, but nothing came of it ; the contact with American business men had suggested openings for commercial enterprise which, later, led to disappointment ; but for the time the main satisfaction was that the western visit was over, and a long holiday at Sheringham with the children occupied all the remainder of the summer. As an example of the spring and vivacity of his nature, after all he had come through, a thoroughly characteristic incident may be given in Lady Shackleton's words :

"One afternoon when in my room at Sheringham I heard that 'an old gentleman, an uncle of Sir Ernest's,' had called to see me. I went into the drawing-room and saw a shabbily-dressed, bent old man, with a grey beard and hair, who gruffly said 'How d'you do?' His appearance was so unprepossessing that I could hardly believe he was a brother of my dear, distinguished-looking father-in-law. But I shook hands politely, summoned up courage to look him in the eyes—it was Ernest! He had gone to Clarkson's in London to be 'made up,' entering into the part so thoroughly that he asked a porter to help him along the platform and into the train; his only regret was that he had to throw away his cigarette, as it made his moustache come unstuck.

"The children were on the sands and we went to see them first, 'Uncle' hobbling along on my arm; then to the golf-links to lie in wait for various relatives, who were as completely taken-in as I had been."

One can imagine the light-hearted laughter. The cheery spirit lasted through an autumn lecturing tour in Scotland, which was superintended by Mr. A. C. Wade, who travelled with him and took a sympathetic pride in the lectures and in the lecturer's personality. His recollections of the tour are full of characteristic touches; we give a sample :

"We visited, together, the fashionable health resorts in Scotland, which at that season of the year are filled with notabilities from all parts of the world, and everywhere Sir Ernest was greeted by large and distinguished audiences. The three weeks' tour he looked upon as a happy holiday. He played golf on some of the most famous Scottish links, which he greatly enjoyed. He was full of life, and always playing practical jokes. He had a keen sense of humour, and had he not been a great explorer, given the opportunity, I think he might have made an equally great actor. One evening, coming off the platform, after having given an exceptionally successful lecture, he rushed up to me in the sideroom, clutched me by the shoulder with one hand, and with eyes ablaze with passion, said, 'Wade! did you see that man with the heavy drooping moustache in the third row of the Stalls—that man is my enemy. I expected every minute he would contradict something I had said.' Then, mysteriously taking out of the tail-pocket of his dress-coat a small revolver, which he had evidently carefully covered with his handkerchief, so that only the muzzle could be seen, said, 'Ah! but I was ready for him, had he done so, I would have shot the beggar at sight.' Naturally I was somewhat alarmed at this sudden outburst, which put Sir Ernest in the best of spirits for the rest of the evening at the success of his melodramatic acting. Some nights later, a friend suggested I should join him in a cigarette, which he fired at me from a toy pistol; then it was I recognized Shackleton's new plaything—the deadly revolver with which he was going so ruthlessly to annihilate his imaginary enemy."

The year included yet one more big lecture tour on the Continent. Leaving Lady Shackleton in a furnished house at Sheringham which rejoiced in the name of Mainsail-Haul, reminiscent of the sea, Shackleton spent the whole month of November in Germany and the adjoining countries, delivering twenty-five lectures in German, of which he had now acquired a practical mastery. There was much travelling, for he visited all the towns of importance from the Rhine to Posen and Königsberg, and in addition Vienna, Prague, and Gratz in Austria, and Bâle and Zurich in Switzerland. Many of the audiences were large and some were enthusiastic; but on the whole the results were less satisfactory than in the spring tour. Shackleton noted a distinct change in the German attitude towards British visitors, and his letters to his wife show that he was often working against the handicap of an uncongenial though scarcely unfriendly environment. He was tired of the trip long before it was done, and thoroughly glad to be home again early in December. It had been a tremendous year of travelling, talking, and receiving honours and hospitality; but the golden harvest of the lectures fell far short of expectation, and the light of common day was coming in as the year of glory was fading away.

The public was never disappointed in Shackleton, and the impression of his gifts as a lecturer left on keen and dispassionate observers may be judged by the words of his agent, Mr. Gerald Christy:

"In the course of my business career it has been my good fortune to become acquainted with all the great explorers, and I am happy to say that they became my personal friends. Shackleton was one of the most appreciative, kindly, and considerate of men. I knew his temperament well. At times he could be a little hasty; and when he was so, that wonderful command of English that was his had a chance to evince itself! But his hastiness made the man more lovable, because the very next instant the fear that he had given offence showed in a contrition that was touching in its way. I vividly remember the square-set, sturdy, forceful figure that he was when he came to see me with reference to accepting engagements for lectures. He was diffident as to his powers as a speaker, but he said that if he had the chance he felt tolerably confident that practice would see him through. It did; and as a lecturer I doubt if Ernest Shackleton has ever been excelled He went from town to town in all parts of the British Isles and to the big cities on the Continent. Not an audience was addressed by him that he did not thrill and amaze and from which he did not evoke admiration, not for himself, but for the band of heroes he felt it so much a privilege to lead. He had a descriptive power without parallel on the platform—at any rate, so I thought."

Mr. Lee Keedick, from his observations in America, corroborated this in all points and added:

"During my whole experience as a lecture manager I have never known any one who could so quickly win the confidence and arouse the interest of newspaper men as Sir Ernest did. The newspapers would often send poets and other literary men to interview him, and although at first they may have taken only a mild interest in the Antarctic, he made such an impression upon them that they would leave thoroughly convinced that Antarctic exploration was the most splendid adventure in the world."

BOOK III
BAFFLEMENT

"Then, welcome each rebuff
 That turns earth's smoothness rough,
Each sting that bids nor sit nor stand but go!
 Be our joys three-parts pain!
 Strive, and hold cheap the strain;
Learn, nor account the pang; dare, never grudge the throe.

 For thence,—a paradox
 Which comforts while it mocks,—
Shall life succeed in that it seems to fail:
 What I aspired to be,
 And was not, comforts me:
A brute I might have been, but would not sink i' the scale."
                                                ROBERT BROWNING

CHAPTER I

UNREST. 1911-1913

"I cannot rest from travel. I will drink
Life to the lees   all times I have enjoy'd
Greatly, have suffer'd greatly, both with those
That loved me, and alone; on shore and when
Thro' scudding drifts the rainy Hyades
Vext the dim sea: I am become a name . . .
How dull it is to pause, to make an end,
To rust, unburnish'd, not to shine in use!
As tho' to breathe were life."

TENNYSON.

THE world's applause had been pleasant, but the year of lecture tours had not proved the financial success which Shackleton's fond hopes had pictured it. His expenses had been very heavy, his liberality had been unchecked by prudence, and at the end of 1910 he found himself no nearer to wealth than when he left the *Nimrod*. With his opportunities another man might easily have laid by the nucleus of permanent comfort; but would another man have made his opportunities?

We need not enter fully into the various business ventures into which Shackleton poured the whole force of his personality for nearly three years. They were all fine, attractive adventures, but their aim was a pecuniary return large enough to ensure the future of himself and his family in luxury, and that they did not achieve. It may be that a man essentially fitted to be an explorer is necessarily lacking in the instincts which lead to success in business, and is attracted more by the potentiality of wealth or of usefulness in schemes than by the prosaic probabilities on which sober men of business build up a fortune. It may be that he could not select subordinates or associates in business with the sure insight which had served him so well in exploration. There may be many other reasons, but the fact is that Shackleton had the soul of a poet, not of a trader; he could rise to the height of any great idea, but he could never sink to the important littlenesses of method and routine which count for so much in business He was often ill-advised and ill-served, but he never tried to take an unfair advantage, and he always resisted the continual solicitations of promoters of speculative public companies to allow his name to appear in their lists of directors In many ways Sir Ernest Shackleton in the city was like Ernest Shackleton on the *Hoghton Tower*, struggling with more moral successes than failures against a strange and uncongenial environment, and the experience had its effect on his subsequent career.

In the autumn of 1910 Scott set out in the *Terra Nova* with the largest and best equipped expedition which ever left for the Antarctic, and Shackleton gave him all the help that his old leader desired Amundsen sailed in the *Fram* ostensibly for a five-years' drift which was to take him across the North Pole from Bering Strait, but at Madeira, where he called as every one believed on his way to round Cape Horn, he electrified the world and his own men by announcing that his aim was the South Pole.

Bruce had issued the prospectus of an expedition to cross the Antarctic continent from the Weddell Sea to the Ross Sea, taking the South Pole by the way; and in Germany, Lieutenant Filchner also had a plan well in hand for an expedition to the interior of Antarctica by way of the Weddell Sea; moreover, the Japanese Lieutenant Shirase was known to be on the point of starting for King Edward Land. There had never been a time when so many aspirants to polar honours were making simultaneously for the far south, and Shackleton felt it terribly hard to stay at home. He could not go out in a subordinate capacity, so he controlled his soul to patience and resolved that in a couple of years, when the gold mine concession he was hunting for in Hungary had made him rich, he would set out once more.

In the early months of 1911, Shackleton was often on the Continent on business enterprises, giving a lecture occasionally in German, in which he was now proficient ; and backing up Filchner on the platform with appeals that greatly helped that young adventurer in getting his funds together. Later in the spring, Dr. Douglas Mawson and Capt. J. K. Davis, old comrades on the *Nimrod*, came to England to seek funds in support of the Australian expedition to the lands discovered south of Australia seventy years before by Dumont D'Urville and Wilkes, and Shackleton spared time from his exacting business engagements to help them with advice and introductions. When the prospects of British support for the Australian expedition were at their darkest, Shackleton wrote a strong letter to a London daily paper, setting forth the merits of the expedition and the tried ability of its leader ; this occurred at a favourable moment, and contributions came in which ensured the future of the expedition and the purchase of the fine Dundee whaler *Aurora*, a sister ship of the *Terra Nova*.

The Shackletons had now decided to take a house in London, and they were busy in seeing to its decoration and furnishing for several months, a work into which Shackleton threw himself with characteristic energy and thoroughness. He was fastidious in matters of taste and finish, and was always planning little surprises for his wife, such as laying parquet surrounds to set off the drawing-room carpet, or designing some delicate harmony of paper and paint. They settled into 7 Heathview Gardens, Putney Heath, in April, when the outlook over Roehampton Common was at its loveliest in the spring of an exceptionally beautiful year.

No one appeared happier or more contented with his lot, but the voice of the wild was always calling, and when he stopped to think the call grew strong. Still, for a time it was stifled. His old school-friend, Mr. John Q. Rowett, allowed him to use his office as a city address, and there was much going to and fro between London and Budapest. He started on one of these journeys on the morning of 15th July, expecting to be back in ten days, a fortnight before anything of the kind was expected, but was met at Boulogne by a telegram announcing the birth of his second son. The business was at a critical stage and he had to go on. Nothing came of it, though that was not known for some months, but the Hungarian friend who was acting with him, Mr. Sandor von Hegedus, became godfather to the newborn Edward. The home-life of the family at Putney Heath or in summer quarters at Eastbourne filled a large place in his life, and he was great in the invention of "surprises," sparing no trouble when the season came round to get himself up as the most realistic Father Christmas who ever came through a snowstorm to waken small children in their beds on Christmas Eve.

Early in 1912 Shackleton's thoughts were turned strongly to the south. He pictured Amundsen and Scott at the Pole. He pictured Mawson making his way to Adélie Land in the *Aurora* and Filchner forcing a passage into the Weddell Sea in the *Deutschland*, and the thoughts were envious, we fear. In a letter to a great friend in New Zealand, dated 12th January, he says :

"I owe you a letter for a long time, but somehow I am always scuttling about the country and get little time to write. I wish I could get another Expedition and be away from all business worries. All the troubles of the South are nothing to day after day of business. My wife and the three children are well I see little of them, though. I suppose we shall soon hear of Scott I am inclined to think that we will hear from Amundsen first. I am looking forward to news."

On 9th March the news arrived that Amundsen had reached the South Pole on 14th December 1911, and no one paid a more generous tribute to the successful explorer than did Shackleton, who had failed to anticipate him in 1909 by so narrow a margin. In an article in a newspaper he says :

"The outstanding feature, to my mind, of the whole of this great journey is that Amundsen made for himself an entirely new route. Leaving his winter quarters he pushed south over an unknown part of the Barrier surface, and instead of journeying westward to go up the great Beardmore Glacier, he still kept his sledges pointing due south. . . . The attainment of the geographical South Pole has been the ambition of every

LADY SHACKLETON AND THE CHILDREN, 1914.

explorer, without exception, who ever proposed to make an inland journey towards that point; and to that end, I am convinced, all matters of detailed scientific work would be of secondary consideration. . . . The discovery of the South Pole will not be the end of Antarctic exploration. The next work of importance to be done in the Antarctic is the determination of the whole coast-line of the Antarctic continent, and then a trans-continental journey from sea to sea crossing the pole "

The *Terra Nova* reached New Zealand early in April, and reported that Scott had not returned from the Southern journey when the ship was forced to leave M'Murdo Sound on 7th March.

There was a renewal of public interest in the Antarctic, and every spare evening in Shackleton's busy life was seized upon for lectures. He delivered two in the Guildhall, and the Guild of Freemen presented him with an ornate illuminated address of thanks which was handed to him by the Lady Mayoress. He lectured to various schools and societies, spoke in aid of innumerable charities, opened bazaars and flower-shows, though he protested, to the obvious pleasure of his hearers, that he was " more at home opening a tin of sardines."

He went much into society, cultivating the acquaintance of wealthy business men, and he was so prominent in the public eye that one sometimes heard the remark that Shackleton lost no opportunity of advertising himself. A little incident of this period bearing on his essential modesty was known to no one but himself and his publishers, and has never got into print before. A condensed narrative of the *Nimrod* expedition was brought out at a very low price for use as a reading book in schools ; it appeared with the head of a Greek warrior on the cover and " The Hero Readers " as the name of the series. This was so displeasing to Shackleton, who hated to think that any one should suppose that he was claiming to be a hero, that he insisted on the cover being changed and the fly-leaf with " The Hero Readers " removed before he allowed the edition to be sold.

The disaster to the White Star liner *Titanic*, in which 1500 lives were lost in consequence of the ship striking an iceberg in the North Atlantic, thrilled the world in the spring of 1912 with a horror even greater than that which attended the deliberate atrocities of the Great War, for generations of peace and safety had drawn a veil of forgetfulness over the terrible possibilities of life. Shackleton took a prominent part in the inquiries held into the cause of the disaster, and the best means of preventing similar events in the future. His evidence at the public investigation was clear and forceful. He spoke very strongly as to the danger of running ships at high speed in a fog, and declared that no shipmaster would do so if he did not believe that it was the strong desire of the owner that he should take great risks rather than lose time. He also pointed out that a look-out man on the mast cannot see small icebergs at night even in clear weather as well as if he were stationed close to the water-line, this being a fact that the voyages of the *Discovery*, the *Morning*, and the *Nimrod* had impressed on his mind ineffaceably.

Two months in summer were spent at Seaford, where Sir Ernest and Lady Shackleton played golf diligently, and the strain of life in London was relaxed for a time. He welcomed Amundsen on his visit to London in November; and when the hero of the South Pole read his paper to the Royal Geographical Society, Shackleton proposed the vote of thanks with characteristic heartiness, acknowledging that he would have preferred if the triumph had been achieved by a British explorer, but recognizing the fine qualities which Amundsen and his men had shown in the plan and execution of their journey.

The *ignis fatuus* of Hungarian gold mines had flitted beyond the horizon, but towards the end of the year Shackleton had been persuaded that his long cherished little cigarette business would blossom into a great fortune if it could be successfully transplanted in America. In a few months he would see himself secure for the future, his debts, incurred largely in trying to help others whose misfortunes were greater than his own, could then be paid and leave him free to organize the expedition on which his heart was now firmly set There had been some talk of applying for the position of an Elder Brother of the Trinity House, carrying a salary of £1000 per annum; but it was put aside (not now the days when he longed to be "a captain with £300 a year") He said, "What good would a thousand a year be to us ? " He turned a deaf ear to the urgent request of the Unionist Party organizers that he would stand as a parliamentary candidate once more.

He went to New York in December and had a strenuous time there, living hard night and day in business, and in the restless excitement of the richest and gayest circle of society. Here his popularity was unbounded, and his sensitively sympathetic nature, finding something in common with all that is human, led him into an apparent adaptation to an environment that was really foreign to his deepest feelings.

News of great interest as bearing on his future work came from the far South in the early weeks of 1913. Filchner's expedition had returned to South Georgia after having forced its way through the ice-encumbered Weddell Sea to 77° 48' S., finding a new coast named Prinz Luitpold Land beyond Bruce's Coats Land; but a landing was frustrated and the *Deutschland*, beset in the ice in March, was drifted northward for 264 days, including the whole winter, passing a short distance to the east of the charted position of Morrell Land with no sign of it to be found, and late in November being set free by the ice breaking up in latitude 63° 40' S.

Grievous news came from the other side of the Antarctic continent. On 10th February 1913 the *Terra Nova*, which had gone down to M'Murdo Sound to bring back Scott's expedition, returned to New Zealand and reported that the Southern Party, travelling by the Beardmore Glacier, had reached the Pole in 1912, a month after Amundsen, but on the return Edgar Evans had died on the Glacier and the others had been overtaken by a series of heavy blizzards, and while confined to their tent by bad weather, Scott, Wilson, and Bowers had perished from want of fuel only 11 miles from the depot where plenty awaited them. The circumstances were so closely parallel to those in which Shackleton had so nearly lost his race with Death in 1909 that the contrast in the result could not fail to impress him profoundly.

On his return from America in February, Shackleton lost no opportunity of testifying to his respect and admiration for his old chief, and did his part in helping to raise the tremendous wave of sympathy and sorrow which lifted the British people to an unprecedented height of hero-worship. A large fund was raised in a few months and most liberal provision made for the future of Scott's wife, son, mother, and sisters, and for the dependants of several others who perished on the fatal return from the Pole. It was held then that the heroic fortitude of Scott, Wilson, and Bowers, the noble self-sacrifice of Oates, who walked out to his death in order to increase his comrades' chance of escape, were deeds unheard of in an age of degenerate love of ease and pleasure, and, therefore, worthy of this unique memorial. A few years later the nation learned with amazement that heroism no less magnificent was latent in every British youth, and that at the front, in the Great War, men were being sentenced to death by court-martial if they failed to rise to heights which, seen for the first time after ages of peace and comfort, appeared almost too lofty for human attainment.

Soon came news from the *Aurora* that Frank Wild was safe home from the Queen Mary Land which he discovered, while Mawson remained behind in Adélie Land where two of his companions had perished   But Shackleton was off again to New York in the *Mauretania* for six weeks' strenuous work on his tobacco business, from which he confidently expected an early return of thousands a year. There was little time on this occasion for social dissipation, but enough for this cable to his wife on the anniversary of their wedding day :

" Happy receiving yours. Have not forgotten the day. Expect sail 23rd. Business growing rapidly. Fond love.
" ERNEST."

When he got home the air was still full of Antarctic memories and projects. Davis was in London struggling hard to raise subscriptions to enable him to take the *Aurora* south again for the relief of Mawson ; Wild was in London full of his year in Queen Mary Land, eager to find a place on a new expedition, and as devoted as ever to his old " Boss." Many of the members of the Scott expedition had returned, and Shackleton was closely in touch with them. On 9th June he took the chair at a great meeting in the Queen's Hall, when Commander E. R. G. R. Evans, R N., Scott's second in command, delivered his lecture. Here Shackleton, who had no opportunity when the memorial meetings had been held at the Royal Geographical Society earlier in the year, paid a fine tribute to the memory of his old chief, to his dear old friend Wilson, and to the other heroes of the fatal return journey from the Pole, and he summed up the moral by showing the bright side of human tragedy in the words, " Death is a very small thing and knowledge very great."

In the course of the summer James Murray and G. Marston produced a little book, *Antarctic Days*, recounting some of the lighter aspects of the *Nimrod* expedition, and telling many good stories of life in the hut at Cape Royds, a few of them sly digs at Shackleton himself, who wrote in a cheery preface :

"All polar explorers are optimists with vivid imaginations, and optimists with vivid imaginations lay themselves open to criticism, not to be expected or deserved by those who are ' ta'en in earth's paddock as her prize.' . . . If I had been asked to contribute to the book I certainly would have tried, as the Americans say, ' to get my own back ' on the authors who were my acquaintances in 1907, and who, in 1913—I hope they will agree with me—are my friends."

In such an atmosphere it was impossible for Shackleton to contemplate living on in the pursuit of wealth so elusive as that which always danced before his sanguine vision just beyond his grasp, and he gave some attention, as he put it, to "nursing millionaires who could put down £100,000 if they cared to," for at last in his confidential correspondence, when worried, tired, and depressed, he acknowledged, " I suppose I am really no good for anything but the Antarctic." One wealthy friend was so far interested as to give an Antarctic dinner at a famous restaurant to all the available men who had been in the far South. The table was transformed into a picture of the Antarctic with artificial snow and real ice, where large models of the *Nimrod* and the *Aurora* were placed at the edge of an ice-barrier thickly peopled by penguins, and Marston, the Antarctic

192 THE LIFE OF SIR ERNEST SHACKLETON

artist, painted special menu cards. It was a gay gathering, enlivened towards the close by songs from Harry Lauder; but, alas! the rich and kindly host did not pursue his hospitality to the point of providing transportation for his guests to revisit the haunts they longed for.

Shackleton gave many lectures and assisted at many public functions that summer, and his friends saw plainly that the long period of unrest was drawing to a close, the lure of the Antarctic was overcoming the excitement of city life, and hints as to an impending expedition dropped from him whenever he spoke. He had one happy week-end alone on the Norfolk Broads in a yacht, with his wife and Frank Wild. He was tired of the towns, where he could not feel the freshness of the winds nor see the stars—the blaze of sky signs on the farther bank of the Thames in the year before the war, hid even Orion and left Sirius himself an inconspicuous spark

CHAPTER II

THE *ENDURANCE*. 1913-1915

" Weary of myself, and sick of asking
What I am, and what I ought to be,
At this vessel's prow I stand, which bears me
Forwards, forwards, o'er the starlit sea.
And a look of passionate desire
O'er the sea and to the stars I send ;
' Ye who from my childhood up have calmed me,
Calm me, ah, compose me to the end ! '
Ah ! once more I cried, ' Ye stars, ye waters
On my heart your mighty charm renew ;
Still, still let me as I gaze upon you
Feel my soul becoming vast, like you ! ' "

MATTHEW ARNOLD.

BOTH Poles had been reached, and one of the greatest incitements to geographical exploration had passed away ; but Shackleton had been brooding for years on his scheme for crossing the Antarctic continent from sea to sea, and he knew that it was a big enough adventure to call forth all his powers. He felt, too, that after years of bafflement in business, where every hopeful enterprise led only to worry and disappointment, he must free his soul, returning to his old ideals. His friend Dr. Bruce had not succeeded in getting support for his project of crossing from the Weddell Sea to the Ross Sea, and although he had priority in the published plan, he had not succeeded in raising funds during the six years his scheme had been before the public, and with the generosity of his nature he gave way to Shackleton's greater chance of success. This he did the more willingly because a foreign rival was in the field. Dr. König, with the support of the Vienna Geographical Society and the assistance of Lieut.

194 THE LIFE OF SIR ERNEST SHACKLETON

Filchner, had put forward plans for an oceanographical and exploring expedition to South Georgia and the Weddell Sea under the Austro-Hungarian flag. Neither Bruce nor Shackleton could bear the idea of leaving the field free to a foreigner, and Shackleton used the fear of foreign rivalry to stimulate interest in his expedition as he had done before, and this time with a greater probability of interference from abroad than had befallen him from Belgium when he set out in the *Nimrod*.

The week-end on the Broads with Frank Wild in the summer of 1913 had shown him his second in command ready to join him at a moment's notice, and from that time everything gave place to preparations for a new expedition. Cautious friends dissuaded him, of course. They pointed out that second polar ventures by men who had led one successful expedition had usually resulted in failure, sometimes in disaster. They urged on him, as they had urged on Filchner, the folly of taking elaborate equipment for a land journey into the Weddell Sea where no safe landing place had ever been sighted, and where no two voyagers had found similar conditions of ice to prevail. They suggested that a preliminary expedition for the exploration of the coast by sea should first be made, and a future land expedition based on the results which might be ascertained. But it was too late. Shackleton had made up his mind, and whatever misgivings geographers might have as to the prospects of the venture, there was agreement on one point: if any man could carry through such an expedition, Shackleton was the man who could do it.

He himself had no doubts as to what he ought to do. The future had gripped him, and his confidence and enthusiasm had impressed a wealthy man so deeply that the funds were all but in the bank. Not quite, however, for the promised patron was a Spiritualist who regulated his life by the revelations of a medium, and the medium for month after month advised waiting for light. Meanwhile Shackleton had rented an office at 4 New Burlington Street, and had taken many important steps towards the completion of the plans of the Imperial Trans-Antarctic Expedition. During the summer and autumn, business matters had been arranged so as not to

SIR ERNEST SHACKLETON, AGED 40, IN SLEDGING DRESS.

Face p. 194

interfere much with expedition work; a large sum from one of his enterprises was at last in sight, almost in hand, and as the house at Putney Heath was too far from the centre of activity the Shackletons took a new house, 11 Vicarage Gate, Kensington, confidently expecting that the difference in rent would be more than made up by the reduction in taxicab fares.

The first public announcement was made in a letter to *The Times* on 29th December 1913, when it was stated that the generosity of a friend had made the expedition possible. The immediate result was the offer of their services on the expedition of nearly five thousand people, from whom only fifty could be taken. Soon afterwards a curious friend noted in his office three large drawers labelled respectively " Mad," " Hopeless," and " Possible " in which the letters of application had been roughly classified. The generous friend found that his spiritualistic adviser would not sanction his helping the funds of the expedition, and in the beginning of 1914, when the return of the Australian Antarctic Expedition under Dr. Mawson was reviving public interest in the South Polar regions, Shackleton found himself with arrangements well in progress and no money to go on with. He would have liked to meet with one or two like-minded with himself who would find money or furnish guarantees to enable him to proceed quietly as in the case of the *Nimrod*, but he had only secured two large sums, gifts in his belief, though subsequently repayment was required and had to be made. The situation which now faced him was serious in the extreme. He must either find £50,000 in a few weeks or acknowledge that his expedition was a fiasco. The plan of the expedition was briefly this Shackleton with a small party was to land in the south of Luitpold Land in the Weddell Sea, and march to the South Pole, continuing thence by the Beardmore Glacier to the Ross Sea, where the second ship of the expedition would be in waiting to bring him home. The proposed route is shown on the map on p. 196.

Two ships had been secured. The *Aurora* of Mawson's expedition, which was believed to be in good seaworthy condition, would be available in Australia for the Ross Sea

196 THE LIFE OF SIR ERNEST SHACKLETON

party, and so the expense of the 12,000 mile voyage from England would be eliminated. The ship for the Weddell Sea party had been found in Norway, where she had been built a few years previously for polar tours, with accommodation for ten passengers and special fittings for scientific research.

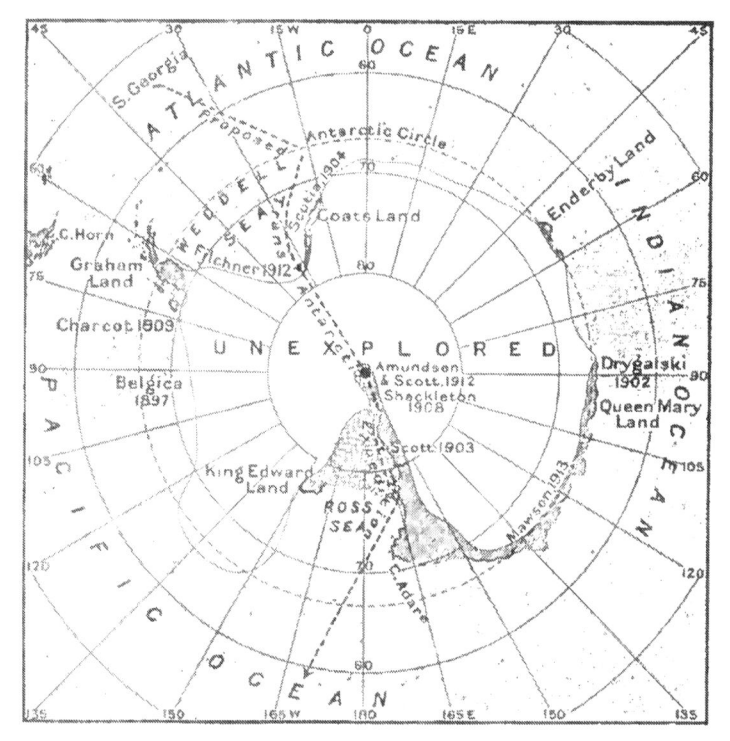

ANTARCTIC REGIONS AS KNOWN AT SHACKLETON'S DEATH, 1922.
(Showing proposed route of Imperial Trans-Antarctic Expedition, 1914.)

She cost £14,000, was named the *Polaris*, and was intermediate in size, between the *Nimrod* and the *Aurora*. It was arranged that she was to be re-named the *Endurance*, a reference to Sir Ernest's family motto, "By endurance we conquer." A hundred big Canadian sledge-dogs were being selected in the far north of Canada, and a new motor sledge with a propeller similar to that of an aeroplane had been designed, as well as new tents more roomy and easier to erect than the old patterns, new rations, and a host of improvements in equipment.

With all these commitments and the growing insistence of Dr. König that the Weddell Sea was to be reserved as an Austrian lake for the next Antarctic summer, the position of Shackleton had become very difficult. He was tired with months of intensely hard work, harassed by the illness of his children, and saddened by the death of his wife's kindly brother, with whom he had spent so many quiet and happy week-ends at Hinton Charterhouse in the great days of popularity.

He rose to the difficulties as of old. In a very short time an attractively printed prospectus of the proposed expedition, in the form of a large quarto pamphlet, was got ready and a list compiled of several hundred wealthy people whose record suggested that they might be of the order of intelligence which could appreciate great enterprises, aimed not at gain, but at glory and the honour of the flag. The result was eminently satisfactory. Shackleton wrote a personal letter with each pamphlet, asking for a subscription of £50 to the funds, and offering an interview to explain his designs further. In this way he secured a sufficiency of money to allow the expedition to go forward, and, what he appreciated no less highly, he made several new friends who could understand his ideals as well as his ambitions, and who valued his friendship long after the dashing adventure had come to naught. Mr. Robert Donald, then editor of the *Daily Chronicle*, supported him with advice and introductions that were of the utmost value.

The great Dundee jute manufacturer, Sir James Caird, whose austere simplicity of life would not encourage a visitor to expect much in the way of pecuniary assistance, asked for an interview with Sir Ernest and questioned him keenly, not only on his plans but on his financial arrangements, laying it down that he would give no help unless the expedition started clear of debt. Shackleton told him that his future book and lectures had been pledged as security for the advances promised by certain guarantors, and Sir James said very quietly and with great deliberation, " Do you think, Sir Ernest, that those gentlemen would release you from that obligation if you were able to tell them that there was a man in Scotland who would find the remaining twenty-four thousand pounds on that condition ? " Shackleton nearly fell off his chair with the shock of surprise and relief, for here at a stroke was an end to all his anxieties. He undertook to see that the condition was fulfilled, and left Dundee not this time a rejected candidate, but an explorer set free.

The promise was confirmed and the cheque sent in the following letter.

"DEAR SIR ERNEST SHACKLETON,—The account you gave me of your plan of going from sea to sea is so interesting, I have pleasure in giving you my cheque for £24,000 without any conditions, in the hope that others may make their gifts for this imperial journey also free of all conditions —I am, yours truly, JAMES CAIRD."

Miss Janet Stancomb-Wills (now Dame Janet Stancomb-Wills, D.B.E.) also gave generously to the funds, and took the deepest personal interest in the expedition. Shackleton was always responsive to the sympathy of understanding women, and in the difficult days before he left England to join the *Endurance*, he and Lady Shackleton were frequently the guests of Miss Stancomb-Wills, whose kindness to the family never failed in the long period of suspense and anxiety which followed. Miss Elizabeth Dawson Lambton, who had helped the *Discovery* and the *Nimrod*, was no less interested in the *Endurance*, and never faltered in her belief in the leader The financial aspect of the preparations must not be left without recording that the Government gave a grant of £10,000 and the Royal Geographical Society, although its resources had been crippled by large donations to other expeditions and by its removal to new premises, testified its good will by a grant of £1000.

In May Shackleton spent some weeks in Norway with half a dozen of his men, camping on a glacier at Finse near Bergen, testing the new pattern of tents and the new motor sledges. Unfortunately, although the ice and snow were present in full polar difficulty, there was the complication of heavy rain, which is unknown in the Antarctic regions, and with which the gear was not designed to cope.

At home the work of equipment was going on apace. The personnel of the expedition was completed early in the summer. Captain Æneas Mackintosh was to command the *Aurora* and lead the Ross Sea party, which was to lay out a line of depots to the Beardmore Glacier, to assist the trans-continental party, and then to winter at M'Murdo Sound, keeping the *Aurora* there in winter quarters. The *Endurance* found a skilled navigator with experience of the Newfoundland ice in Captain Frank A. Worsley, whose cheery personality and gay contempt of danger fitted him well for the post. Several officers in the Army who were tired of the dull inaction of their profession, and anxious to see danger and excitement, secured places on the expedition. Wild, of course, was on the spot, and two other veterans of the *Discovery*, Crean and Cheetham (who had been one of the officers of the *Nimrod*), joined as second and third officers. Most of the others were new to the Antarctic, but all were picked men with fine records and that passion for adventure which the leader recognized as the prime requisite for every member of the expedition.

Everything was at last beginning to go well, and Shackleton went here and there through the country delivering lectures to keep up the public interest, and every now and again to help a charity. He was appointed President of the Browning Settlement at Walworth, where he assured the assembled working men of the help which Browning's cheery optimism had always been to him, and promised to name a mountain after the poet if he should discover new peaks in the forthcoming expedition.

At Glasgow in June he received the honorary degree of LL.D. from the University, and at a dinner in the evening was called on to reply for the new graduates. Part of the proceedings of the day had been a memorial lecture on Lord Lister, and Shackleton referred to the fact that Mt. Lister was the loftiest summit yet found on the Antarctic continent, and that the great surgeon had appreciated the compliment of placing his name so high in that absolutely germ-free atmosphere. Lord Rosebery, who followed, challenged the statement that no germ was to be found in the pure Antarctic air; on the contrary, he said that he believed that analysis would discover a germ, infection with which had compelled Sir Ernest to do heroic deeds and lead a pleasant life in darkness and the society of penguins. He could not think that anything less than a bacillus of the most virulent and persistent nature could account for a man being willing to face such hard conditions a second time.

About this time the business control of the expedition was confided to the care of an eminent firm of solicitors, and Mr. Alfred Hutchison, one of the partners, threw himself heart and soul into the work, and remained a trusted and helpful friend of Sir Ernest to the end.

The *Endurance* reached the Thames in June, and stores were being put on board; the hundred Canadian dogs were recovering from their long journey in a rural retreat near by, and Shackleton was getting the last of his scientific staff together, electrifying the young University men who were going with him by the rapidity with which he judged their fitness, and the short time he allowed them to gather the necessary apparatus for their respective departments In truth, we must allow, that vastly as Shackleton's expeditions have advanced several branches of science, he himself had little sympathy with scientific methods, and could never understand the necessity of long preparation for scientific research in new conditions. But he was determined that his expedition should have a good scientific record to its name, so he chose good men, young, full of ambition and determination, and left them a pretty free hand to do their best. The method was rather like teaching a puppy to swim by throwing it into the water, and at first it took away the breath of the academic youths who had been accustomed to have their way prepared and watched over by a professorial providence. Later they responded to the method, and were not second to the old Discoveries and Nimrods in devotion to " The Boss."

The labour of preparation increased as the date fixed for sailing approached. It was arranged that the *Endurance* should leave the Thames in time to be at Cowes in the course of the Regatta week for a final inspection and farewell by the King, as his father had inspected and bidden farewell to the *Discovery* and the *Nimrod*. A good deal of equipment had still to be delivered from Norway, and several hundred pounds worth of scientific apparatus had yet to come from Berlin and Paris; but the main bulk of the stores was at the dock-side or stowed on board by the middle of July.

On 16th July Queen Alexandra, whose interest in Antarctic expeditions had not abated, paid a visit to the *Endurance* in the South-West India Dock accompanied by the Empress Marie of Russia and Princess Victoria. Her Majesty inquired after the hundred dogs which at first she was keen to visit, but on hearing that they were doomed never to return, she drew back declaring that she could not bear to look at the poor creatures. She inspected every part of the ship, and presented various souvenirs, including a large Bible with an autograph inscription for the crew, another for the wintering party, a miniature of the Queen Mother's Royal Standard, and an enamel medallion showing St. Christopher the patron saint of ferrymen. When in due time the ship sailed, Queen Alexandra sent a touching telegram of farewell and God-speed.

Dr. König was pushing on with his expedition and still demanding that he should have the Weddell Sea to himself, and Lieutenant Filchner was still trying to bring the two expeditions to some sort of *modus vivendi*. He wrote begging Shackleton to come to Berlin in the last week of July to meet König, but Shackleton could not leave his own preparations, now in the agony of completion, and suggested that König should come to London. Nor had he had time to follow the darkening of the political firmament, or if he did it would only have been to think like others that storms so terrific as those clouds portended did not occur in our civilization. Had he gone to Berlin there is little doubt that his endurance would have been tested for the next few years, not in the southern ice, but in the Ruhleben internment camp.

The *Endurance* left London on Saturday, 1st August, flying the blue ensign with the burgee of the Royal Clyde Yacht Club, and loaded with mascots to secure a lucky voyage. She drew away from the wharf to the sound of a pibroch discoursed by a Highland piper in compliment to the Scottish contribution in money and men. The jollity of the scene was superficial so far as the leader was concerned. The war-cloud had burst over Europe, and though no more was heard of the König Expedition, the universal mobilization on the Continent had created a most difficult situation. The delivery of the scientific apparatus from Germany was hopeless, and that from France was also stopped. Now the King telegraphed that he could not go to Cowes this year. The *Endurance* lay off Margate for Sunday, and on Monday morning Shackleton went on shore and saw the order for general mobilization in the newspapers. Several officers who were ready to sail left at once to rejoin the colours. Shackleton returned to the ship, mustered all hands and told them that he had decided to place the ship, crew, and stores at the disposal of the Admiralty for service in case of war. Every one agreed and a telegram conveying the offer was sent to the Admiralty, and in an hour came the answer, contemptuously laconic—so it seemed to the patriotic ship's party—" Proceed." An hour later, a long and courteous telegram from the First Lord, Mr. Winston Churchill, arrived, thanking the commander and crew for their generous offer, but saying that as the expedition had been organized with the sanction and support of the highest geographical authorities, the Admiralty felt that it ought not to be interfered with The ship sailed for Plymouth, but stopped at Eastbourne to land Shackleton, who had been in a state of great mental strain in the effort to convince himself as to the right course to pursue. The King received him on the afternoon of 4th August, thanked him for his offer to abandon the expedition, and assured him that it might proceed with his approval and good wishes. His Majesty then presented a silk Union Jack to carry on the trans-Antarctic journey Shackleton concurred, but was not happy.

The ultimatum to Germany expired at midnight, and we were at war. Every one declared that it would be a short war; the conditions of modern warfare were such that victory in six weeks (optimists said six months) or submission through starvation must result. In this hope the *Endurance* sailed from Plymouth at the end of the week, but Shackleton and most of the scientific staff remained to try to get together the essential things which had not been delivered in time. A consignment from Norway lay buried under a mountain of war stores at one of the docks and required heart-breaking efforts to overcome the official and physical obstacles before it could be discovered. The promised free passages and free freights to South America and Australia were suddenly cancelled; the sailings of steamers were altered and could not be relied on, the blackness of the war prospects grew more serious day by day. It was a strain that few could stand, and Shackleton found relaxation in rare visits to his family, then at Eastbourne, where for an occasional afternoon he became a boy again, larking with Ray and Cecily on the rocks and in the water. This gave relief for the time, but more than once Shackleton, who kept up his brave show of optimism to the reporters, acknowledged in confidence that he was almost "on the point of chucking the expedition and applying to Kitchener for a job." But just as on the *Nimrod* off King Edward Land, the thought of responsibility to those who had embarked on his venture and to those who had found the funds for it, overcame his personal feelings. Before he quelled the turbulence of his thoughts he and Lady Shackleton journeyed to Scotland and conferred with Sir James Caird, whose advice helped to overcome his wish to serve his country in the war and decided him to go on with the expedition; but it had been a severe struggle.

At last he got away from Liverpool on 25th September bound for Buenos Aires to join his ship He was utterly tired out, his nerves on edge, and burdened with the new troubles that the war was heaping in his path. He wrote home:

"I love the fight and when things are easy I hate it, though when things are wrong I get worried. . . . I feel that I am going to do the job this time. . . . I don't think I will ever go on a long expedition again. I shall be too old."

The *Endurance* had been leaking on the voyage out, some of the crew had proved unsatisfactory, and it was difficult to supply their place. There were difficulties as to the supply of stores, and the port authorities had not been disposed to put themselves about to expedite the departure of the ship. The officers had done their best through the proper official channels, but they were lost in the land of *mañana*. When Shackleton arrived all was changed. He went at once to the highest authority, found access and stated the case. The charm of his personal appeal had been felt by hard-headed, cold-blooded business men at home ; to an Argentine official steeped in Spanish idealism it was absolutely irresistible. All doors opened, all wheels ran smoothly, everything was done, done quickly, and with an air of Castilian grace. The countrymen of Cervantes recognized and did honour to the spirit of the caballero of La Mancha in this ingratiating and persistent stranger. So on 26th October the *Endurance*, with the Boss and all her complement of men and dogs, departed from the River Plate and in due time dropped anchor off the Norwegian whaling station of Grytviken in South Georgia. Here Shackleton stayed for a month making his final arrangements and recovering from a sharp attack of what he called influenza. The grim snow-topped mountain ridge that was destined to try his mettle and hold his memory was daily in his view, and his mind was full of forebodings all unsuspected by his comrades. He wrote :

"Except as an explorer I am no good at anything. . . . I want to see the whole family comfortably settled and then coil up my ropes and rest. I think nothing of the world and the public. They cheer you one minute and howl you down the next. It is what one is oneself and what one makes of one's life that matters."

The whalers reported very bad ice conditions in the Weddell Sea, and after the opportunity he had had of studying his companions on the voyage across the South Atlantic, Shackleton decided not to send the ship back from her farthest south point, but to seek winter quarters for her and postpone his great land journey until the following year. This decision was sent home in time to reach the *Aurora* before she sailed for the Ross Sea.

Meanwhile a new and unsuspected danger was drawing near ; but it passed. The German Admiral von Spee, with his cruiser squadron, had taken a British collier after the battle of Coronel and was transferring the coal cargo to the warships in Beagle Channel near Cape Horn, whence the squadron sailed on 6th December to its doom at the Battle of the Falklands. Had the coaling been quicker, and had the rumour of a British fleet at the Falklands reached von Spee, it is not improbable that he might have taken refuge at South Georgia and nipped the Imperial Trans-Antarctic Expedition in the bud. As things fell out the *Endurance* left Grytviken on 5th December, and with her convoy of observant albatrosses she was making her way through the South Sandwich Islands when, a few hundred miles distant, Admiral Sturdee was sinking the German cruisers on 8th December.

The company on board the *Endurance* numbered twenty-eight all told. In addition to the leader, five had had previous Antarctic experience : these were Frank Wild, T. Crean, second officer, and A. Cheetham, third officer, all of whom had been on the *Discovery* and several later expeditions ; G. Marston, the artist, who had been on the *Nimrod*, and F. Hurley, the photographer, who had been with Mawson in the *Aurora*. There were also Frank A. Worsley, the skipper ; H Hudson, navigating officer ; L. Greenstreet, first officer ; L. Rickinson, chief engineer ; A. Kerr, second engineer ; J. A. McIlroy and A. H. Macklin, surgeons and members of the scientific staff, together with R. S. Clark, biologist ; L. D. A. Hussey, meteorologist ; J. M. Wordie, geologist ; R. W. James, physicist (the last two being Cambridge men), and T. Orde-Lees, motor expert. The crew also included ten men of various ratings. The original intention was to divide up the party into the ship's company and the land party.

The sea was covered with dense pack-ice as far north as latitude 58° 30′ S., and the southward advance was checked in this temperate latitude, for the year was altogether abnormal; summer seemed to have dropped out of the calendar. The attempt to enter the pack in longitude 22° W. failed, and the *Endurance* coasted the edge of the ice to the eastward as so many of her predecessors had done in the Weddell Quadrant of the Antarctic. On 11th December a lead was found between the floes in 18° 22' W., but so cold was the season that even then, a fortnight before midsummer, it was frozen over with new ice through which the powerful triple-expansion engines forced the steel-clad bows of the stout ship. Progress was slow ; the ice grew heavier from day to day, and the thickness of the floes exceeded anything that any one on board had observed in their many voyages through the Ross Sea. Christmas came, was celebrated with the usual jollity, though without relieving the anxiety that was beginning to weigh on Shackleton's mind, and not until the last day of the year was the Antarctic Circle reached and crossed. The *Endurance* had forced her way into the pack for 480 miles, but it had taken her twenty days to do so, an average speed of only 1 mile per hour, towards the south, though the twists and windings of the way made the distance travelled much greater.

On 8th Januray 1915 the ship cleared the pack in 70° S. and for 100 miles she steamed through open water, passing 500 great icebergs in a single day. On the 10th the dull snow-shrouded heights of Coats Land were sighted, and the course continued parallel with the cliffs of an ice-barrier that projected from the land. The ship was often within a mile of the barrier ice, and soundings showed depths of less than 100 fathoms, so that, although no bare ground was seen, land was certainly very near. The biologist was happy examining the animals and sea-plants brought up in the dredge from the shallow water or caught in the muslin tow-nets, and the geologist had congenial employment also, for he found many interesting specimens of rock in the dredges, the material having been carried down by glaciers from strata deeply buried under snow and dropped on the sea-bed as the warm sea-water melted the ice. Penguins and seals were all round, and the newcomers to the Antarctic were in the high spirits which always result from the attainment of a free field of work in virgin wilds, after weeks of baffling struggle through the ice The leader's eyes were fixed on the horizon ahead watching for the ice-blink that would proclaim fresh troubles, hoping that the open water along the Barrier would continue unencumbered.

They passed Bruce's farthest of 1904, and the land on their left-hand was new, though it was so thickly buried in snow that its features were hidden from sight like furniture under a dust-sheet. Still, new land was there, and Shackleton named it the Caird Coast in honour of his chief supporter. On the 15th a great glacier or overflow of inland ice was sighted projecting far into the sea and forming a sheltered bay where the ship could lie alongside a firm ice-foot like a natural quay, whence easy snow slopes led to the summit of the Barrier. This was an ideal landing-place; but it was only in 75° S. and nearly 200 miles farther from the Pole than Filchner's Vahsel Bay which Shackleton was making for; so it was passed by, an opportunity offered too soon that was never to be offered again. The ice over the land was now seen clearly to rise to heights of 1000 to 2000 feet, some miles back from the sea. Good progress had been made to the southward, as much as 124 miles in a single day, taking the ship beyond 76° S. The ice-conditions became difficult again on the 16th, and a great gale sprang up from the eastward so that the *Endurance* had to make fast on the lee-side of a stranded iceberg, while the smaller bergs and masses of floe ice were being hurried past her by the storm. Next day the ship proceeded under sail, feeling her way along towards the south-west between the floes, now and again forcing a passage through sticky stretches of mixed ice and snow, turning now to the right and now to the left as openings presented themselves.

Just in the same way the *Nimrod* had been nosing her way toward King Edward Land seven years before, always taking the turns that led to safety and finally turning back just before it was too late. Now on 19th January 1915, whether the directing mind was a shade slower in its working, or the signs of danger or safety a shade less obvious than they were before, or whether the hand of Fate seized the wheel of the *Endurance*, she took a turn and was held up by the ice: and she did not turn back in time. The ship was beset in 76° 34′ S. and longitude 31° 30′ W. This was not serious, for there was still a good month of the summer to come, and a strong wind might break up the pack and set her free in a few hours. Moreover, it was soon found that the pack and the ship with it was drifting steadily to the south-westward, the direction in which they wished to go; there was no cause for alarm at all.

On the tenth day of besetment the boiler fires were allowed to go out to save fuel. On one occasion open water appeared not far off, and a desperate effort was made to break up the surface of a frozen lead and get the ship into it, but the effort was vain. One day the "motor-crawler and warper" was got out on the floe and tried successfully. This was a form of motor adapted for getting over rough ice, something after the fashion of the warlike engine familiarized to people at home by the name of tank, provided with winding gear by which, when the motor was anchored in the ice, a sledge could be hauled up to it by a long steel wire wound on a drum. It worked well, and might be useful yet, for the ship was drifting south. So things went on for a month. Then on 22nd February (the anniversary of the *Nimrod's* departure from Cape Royds after landing the shore party, seven years before), a latitude of 77° S. was recorded in 53° W. This was off Luitpold Land, 60 miles from Vahsel Bay, at which Shackleton had been aiming; but the ship was fixed in the floe like a castle on an island, and there was no power to move it towards the land nor any possibility of carrying the vast quantity of stores across the rough ice. Nothing could be done except to regret the lost opportunity of Glacier Bay, and that did not help. Shackleton was disappointed, but not cast down, for he had decided at South Georgia not to attempt the great land journey that summer, and it was easier to winter on the ship than in a hut. He recognized that the *Endurance* was fast for the winter, and hoped that she would remain in the high latitude she had reached. Then all would be well for the next summer. The boilers were emptied to avoid damage by freezing. A ring of dog-kennels was built on the floe around the ship where the animals might have more room, and their human companions be relieved from their rather objectionable proximity.

The cold was now severe, yet the floe was always moving, but its trend was to the north. By the end of April they were a whole degree, or 60 geographical miles, north of their farthest south point, and the ice-floe was driving towards a distant

THE *ENDURANCE* BESET IN WEDDELL SEA, AUGUST 1915.

grounded berg against which the floating ice could be seen ridging and piling up ; there was some anxiety as to what would happen to the ship if she came too near, but she was gradually carried past the danger. The sun should have set for the last time near the beginning of May, but it came back again once or twice, its image being raised above the horizon by mirages which betokened open water and warmer air not very far away.

The winter wore on, the ship drifted hither and thither with the floe, in the main northerly, sometimes faster (once she did 37 miles in three days), sometimes slower, and on 1st July she was in 74° S., 180 miles from her farthest south, and by the end of August she had reached 70° S., or 240 miles farther north. The sun had returned on 26th July. During all this time Shackleton was anxiously endeavouring to keep up the spirits of the men, and it was not very difficult, for he kept them busy, well-fed, and comfortable. The electric light had been rigged up; an Antarctic Derby had taken place on the smooth surface of a frozen lead in which rival dog-teams ran close races, and in September seals began to reappear, and there were hunting parties and feasting on fresh meat. But as the spring wore on the state of the ice became worse. It was subject to sudden and alarming movements, sometimes splitting with a loud report and ridging up with a hideous clangour of thrusting and falling masses as the surface yielded to the terrific pressures caused by wind driving one floe against another, or the whole system against the obstruction of the distant land to the westward.

The general track of the drift (see map on p. 236) was parallel to that of the *Deutschland* in 1912, and in August the *Endurance* passed about as far to the west of the charted position of Morrell Land as Filchner had done to the east of it. Soundings showed a depth of 1700 fathoms, practically disproving the existence of land where Morrell had placed it, and where Bruce believed it to be. This was a definite contribution to geography, and the work of sounding and meteorological observations threw much light on the physical conditions of the ice-covered sea.

Week by week the pressures grew more serious, the groaning

and trembling of the ship as she shared in the strains of the imprisoning floes were horrible and alarming. Here and there frost-mist could be seen rising far off like smoke from a prairie fire, where pools of open water were formed by the rending of the floes.

On 18th October the *Endurance* was lifted high above the sea-surface on a rising pressure ridge, and after a shaking was dropped back into a pool of open water. Steam was got up with all speed, the engine gave a few turns ahead and astern, but the hope of escape was thwarted. The pool froze over, the floes all round came together again and on the 24th the ship was caught between three moving masses to which the thinly frozen surface of the pool offered no resistance. One floe pressed against the starboard side, a second against the port quarter, twisting the vessel and starting the planks, while a third floe striking on the port bow rose up over the forecastle, forcing the ship down by the head. She began to leak badly: the pumps were rigged. For two days every man worked his hardest to keep the water down and save the ship. Shackleton says in *South*:

"The pressure-ridges, massive and threatening, testified to the overwhelming nature of the forces that were at work. Huge blocks of ice, weighing many tons, were lifted into the air and tossed aside as other masses rose beneath them. We were helpless intruders in a strange world, our lives dependent upon the play of grim elementary forces that made a mock of our puny efforts. I scarcely dared hope now that the *Endurance* would live, and throughout that anxious day I reviewed again the plans made long before for the sledging journey that we must make in the event of our having to take to the ice. We were ready, as far as forethought could make us, for every contingency Stores, dogs, sledges, and equipment were ready to be moved from the ship at a moment's notice."

The next day was bright and fine, but pandemonium raged in the floe. New pressure-ridges were rising and rushing forward with a roaring noise towards the ship, which strained and cracked. The rudder-post was torn from her by one thrusting floe, the decks broke up and water rushed in. The end had all but come. The *Endurance* could never float again, but spitted on jagged shafts of ice she hung suspended till the floes closed and gripped her fast. On 27th October the pressures began once more, and at 4 p.m. the order was given to abandon the ship; and the emergency stores, the boats, dogs, and men were moved to a hard, unbroken part of the floe near by. Tents were set up on the ice and a camp made. The expedition according to plan was at an end, there could be no trans-Antarctic march, no meeting with the Ross Sea party, no return in the *Endurance* even. The ship had drifted 570 miles towards home in the 281 days, though she had zigzagged through about 1500 miles, since she had been seized in the grip of the ice. Now she was a total wreck, 180 miles from the nearest land to the west and 360 miles from Paulet Island, where the stone hut in which the crew of the *Antarctic* had wintered after the crushing of that vessel in 1903 was still standing, filled with stores deposited there by the Argentine ship *Uruguay* in the same year. The distance by sea to South Georgia was 1000 miles, to the Falklands 1050 miles, and these were the nearest dwelling-places of men. The scientific staff and crew, exhausted by a day of unceasing toil, crept into their sleeping bags in the tents lying on the 6 feet of ice which floated over 8000 feet of ocean depth; but as for the leader, he says in *South*:

"For myself I could not sleep. The destruction and abandonment of the ship was no sudden shock. The disaster had been looming ahead for many months, and I had studied my plans for all contingencies a hundred times. But the thoughts that came to me as I walked up and down in the darkness were not particularly cheerful. The task now was to secure the safety of the party, and to that I must bend my energies and mental power and apply every bit of knowledge that experience of the Antarctic had given me. The task was likely to be long and strenuous, and an ordered mind and a clear programme were essential if we were to come through without loss of life. A man must shape himself to a new mark directly the old one goes to ground."

And as he mused he saw a sudden crack run across the floe through the camp between the tents. He blew a whistle and brought the men tumbling out ; the tents and gear were shifted from the smaller of the two pieces into which the floe had split and re-erected on the larger. The men went back to rest if not to sleep, and Shackleton to his pacings to and fro alone, listening to the shrieking clamour of rising ice ridges and the heart-rending sounds of collapse and destruction in the *Endurance* from the stern of which gleamed an unextinguished light Another spasm of the dying ship broke the connection, and the light went out ; but the soul of Shackleton was enlarged and set free, doubt and anxiety dropped from him, and he gave himself with all his might to the simple, straightforward fight for the safety of his people, putting behind him the shattering of his own ambitions.

CHAPTER III

THE *JAMES CAIRD*. 1915-1916

"There gloom the dark broad seas   My mariners,
Souls that have toil'd, and wrought, and thought with me,
That ever with a frolic welcome took
The thunder and the sunshine. . . .
It may be that the gulfs will wash us down,
It may be we shall touch the Happy Isles
   .   .   .   .
One equal temper of heroic hearts
Made weak by Time and Fate, but strong in Will
To strive, to seek, to find, and not to yield."

TENNYSON.

THE wreck of the *Endurance* was the wreck of all Shackleton's dreams of a second polar triumph, and a new path to fortune through a second book and another series of world lecture tours; but his optimism helped him to stand the shock. The bigness of his nature rose to the measure of his responsibility and triumphed over the slenderness of his resources.

He stood with all his men on a cracked and crumpled ice-floe far within the Antarctic Circle, drifting vaguely on towards warmer seas and dissolution, liable at any time to be torn asunder or thrust up into ridges of splintered ice. No outside help was possible, and the proposition which absorbed his whole attention was how the material saved from the wreck could be used to bring all his men to safety. For the next year the determination that he would preserve his reputation of never having lost a life on his expeditions possessed him to the exclusion of every other ambition. He laid his plans, changing them as the circumstances altered, and took his resolutions, sometimes with and sometimes against the advice of the few trusted friends whom he consulted, for he felt the responsibility to be his own, and alone he bore it.

His first step was to select the most solid portion of the floe, which was found about a mile to the north-west of the wreck, and here he had all essential stores collected and tethering lines set up for the forty-nine dogs. The three boats taken from the ship were mounted on sledges for ready transport and when all was in order on 30th October, the new settlement was named Ocean Camp. A routine was established to keep every one busy, and the "Boss" set the example of that cheery optimism in the value of which he was so convinced a believer The decks of the ship had to be split open to get at the cases which floated up in the flooded holds, and stores had to be saved without selection as they came to hand. The carpenter was kept at work raising the gunwales of the three larger boats and decking them in forward to fit them for heavy seas when transporting all hands in the time that was to come

For three weeks the wreck of the *Endurance* remained in sight, but on 21st November a shout from Shackleton, ever watchful even when another watch was set, " She's going, boys ! " brought all his men out of their tents to see the sickening sight of their once splendid ship diving through the heaving ice, bow first into the depths. Her disappearance was almost a relief ; the dead past burying her dead ceased to cast the shadow of the might-have-been on the living present For a month the routine of Ocean Camp went on, plenty of food, punctual meal hours and a careful grouping of comrades in the various tents helped to keep all cheery ; but the best stimulus to good spirits was the decision which Shackleton took on 20th December to make a march over the ice towards the distant store hut on Paulet Island near Joinville Land.

To lose no time, Christmas Day was celebrated on 22nd December, and as it was impossible to move all the stores saved from the ship, there was a great feast on luxuries, the food-value of which did not justify the labour of hauling them on sledges. This was the last absolutely full meal that any of them was to have for five months to come. The dog-teams were in good condition, the men full of vigour, and the party moved on with two of the boats and sledges loaded with tents, sleeping gear, and food. An advance party hewed a path through the pressure-ridges to let the sledges pass on their course to the west. The work was heavy, the progress slow, and after a week of ceaseless effort astronomical observations showed that the result was only 7 miles advance. At this rate Paulet Island could never be reached, for the floe was carrying them towards the north in its irregular drift far faster than they could move over the rough surface, dragging the heavy boats. So on 29th December Shackleton resolved to form a camp on the ice once more and leave the decision as to the ultimate avenue of escape until he saw in what direction and how far the floe would take them. Opinion amongst the men was divided; but when Shackleton made up his mind as to the right course he did not change it lightly. It was his expedition and he commanded it. So Patience Camp came into being, named after what was to him the hardest virtue. It was only in his fights with Nature that patience appealed to him at all, in his contests with men he preferred quicker ways of enforcing his will.

This was Shackleton's diary for 1st February 1916:

"65° 16½' S., 52° 4' W   No news
S.E. wind, fine weather
Patience,
Patience,
Patience."

Patience Camp was formed a short distance south of the Antarctic Circle; but week by week it was carried northward, crossing the Circle in January 1916, when the summer sun was strong, and with an air temperature of 36° F. the inside of the tents became unendurably hot. The ice was melting everywhere and everything was soaking wet. Seal meat, hitherto the staple food, was sometimes scarce, and as a precaution against its failure all the dogs except two teams were shot. The wind varied in direction and force, and formed almost the only topic of conversation, for the movement of the floe depended on the wind. A week of south-westerly gales moved

Patience Camp 84 miles to the north in six days. The Camp was now in 65° S. where the drift of the *Deutschland*, four years before, had turned sharply to the eastward; a safe trend for a ship still staunch, but one that would have brought utter disaster on a party dependent on small boats, which, once carried to leeward of the near islands in the zone of strong west winds, could never make the land. How keen the anxiety was for the one or two who knew the track of Filchner's imprisoned ship may be imagined.

Every observation of the sun for longitude was waited on with unuttered anxiety while the position was being worked out; but if the chronometer watches kept true—and they did— each observation showed that the northward drift still held. It was clear that when the pack broke up, refuge must be sought on one of the sub-antarctic islands, and if the drift should trend to the westward, Deception Island might be reached in the boats. Whalers might be found there if the season were not too far advanced; in any case they knew that a little sailors' Bethel had been built at the whalers' chief resort, and the carpenter was already in imagination taking sacrilegious liberties with pews and floor-boards with intent to build a little vessel strong enough to cross the ocean to Patagonia or the Falklands. Their own three boats might be wanted soon, so on 2nd February Wild was sent with eighteen men to Ocean Camp, which had been drifting along in the floe parallel to Patience Camp, to bring back the third boat which had been left there. They brought it in safely; but an attempt to return to Ocean Camp next day failed, as leads of water had opened between and it could never be visited again. The *Stancomb-Wills* had been brought home just in time. On 29th February Leap Year Day was celebrated; every occasion for a show of festivity helped to keep up the hearts of the ordinary human beings who were mixed with this goodly company of heroes, and this strange feast-day saw the last of the cocoa. Shackleton tells in *South* how his comrades fared in those trying days Mr. R. W James, a member of the scientific staff, has kindly given us his recollections of the "Boss" himself:

"During this period he was very particular about the meals of the party. He believed in good food and plenty of variety as a specific against discontent, and insisted on everything being as clean and well served as possible. He believed in the maximum amount of civilization possible under the circumstances, and was a stickler for punctuality at meals.

"My own realization of his best qualities came after the crushing of the ship when the party took to the ice. I had the good luck to be one of his tent-mates during the five and a half months' drift in Ocean and Patience Camps, and I think that time would have made me his admirer if nothing else had done so. He admitted to us a day or two after we had abandoned the ship, that he felt an actual relief when the worst came, because he then knew exactly what had to be faced. The whole of his mind then turned to that one problem, of landing the party without a casualty. Not only the main problem, but its details absorbed him. Food, how to get it, how to eke out our slender stock of preserved food to give the greatest variety to the eternal seal; how to keep everyone employed and cheerful, to keep sleeping bags dry, to nip any sign of pessimism in the bud, the best way of keeping the stores ready for an instant shift; all these things and many more occupied his thoughts by day and most of the night. To a man of his temperament the enforced inactivity of those months must have been irksome in the extreme. Yet he had the strength to stop our two attempts at marching when it seemed that nothing could be gained by going on and possibly much lost, and to form a fixed camp the second time contrary to the opinion of many of the party. Yet who can say events did not justify the decision?

"He was an excellent tent mate, and once inside the tent dropped to a very large extent the Commander. We had great discussions about all manner of things. One of his chief arguments was in favour of 'practical' instead of pure scientific research. . . . Sometimes he would be reminiscent, and those times were most enjoyable, for he had met many people from kings down; and he told a tale well, and had a sense of the humour of a situation He would discuss new expeditions, not only polar, but for hidden treasure, and schemes of all kinds for getting rich quickly, and one would realize what a gambler he was. Or he would read or recite poetry, and then one would see quite a different side of him.

"One of our favourite amusements was the game known as 'animal, vegetable, or mineral,' in which one of the players has to guess some object agreed upon by the rest, by asking

218 THE LIFE OF SIR ERNEST SHACKLETON

questions to which the only answer allowed is ' Yes ' or ' No.' Shackleton had quite an uncanny skill at this game. By a few judicious questions he would narrow down the field of inquiry and rapidly arrive at the answer, however remote the thing might be."

By the middle of March the drift had extinguished all hope of trying for Paulet Island, and before the end of the month the first land appeared far to the west, where the mountains of Joinville Island showed up faintly. The chance of making Deception Island vanished soon after, as the ice floe reached the mouth of Bransfield Strait and still held on its northward way The only hope now was to reach one of the two easternmost members of the South Shetland group, Clarence Island or Elephant Island, where no landing had ever been made, and where no succour, save that of solid land beneath one's feet, could be looked for Conditions on the floe were growing worse week by week On 29th March a shower of heavy rain fell, the first experienced since leaving South Georgia, sixteen months before ; it made life supremely miserable, as the whole surface of the camp became slush, inside the tents as well as outside. The last teams of dogs had been shot, as it was feared that, with the approach of winter, seals would disappear. Rations had to be reduced

The little community began to show signs of cleavage from the working of privation and hardship on minds and bodies of unequal resisting power. This constituted Shackleton's greatest preoccupation and received his most vigilant attention. If dissension appeared there would be no chance of safety for all ; divided counsels were not to be tolerated for a moment. He knew intimately the disposition and idiosyncrasies of every man, and day by day the true self of each emerged from the wrappings in which conventional manners disguise character. Roughly the men fell into three classes. Those of experience and sagacity who saw as clearly as Shackleton himself what had to be done, and aided him wholeheartedly in doing it ; those who, with little experience, recognized by a sort of instinct the power and wisdom of the Boss, and were ready to follow him everywhere and trust him absolutely ; and the rest, a few, who, with untrained intelligence or enfeebled bodies, gave way to fear for their own safety, and by brooding on danger had come to the risk of panic or collapse. Those of the first order, chief amongst them Frank Wild, were Shackleton's other self, with whom in absolute confidence he discussed all his plans, and on whom he could rely to the uttermost. Those of the second order had absolute confidence in Shackleton, were sure that whatever he decided was the best, and to them he was always a considerate friend; to the minority he was a master who must be obeyed, and he made himself obeyed and so saved them all.

By the beginning of April the conditions were sufficiently alarming The great massive floe, extending for hundreds of miles, in which the *Endurance* had been crushed and swallowed had now thinned down by melting between the warm sea-water below and the sun above. It had broken into a number of fields of ice an acre or so in extent, on one of which stood Patience Camp with its tents, its blubber stove, its outlook post built up of ice-blocks, its three boats, the *James Caird*, the *Dudley Docker*, and the *Stancomb-Wills*, the solitary remnants of the munificence of those who gave them The swell of the great ocean made the cake of ice heave and tremble; when the wind blew, one or another of the neighbour floes would jog against it and swing back, a crack would follow, and so by mutual approach the ice-fields were breaking up from acres into roods. In the lanes of water between the pieces of floating ice the men could see the killer whales cruising, with their sharp eyes lifted now and then, greedy as sharks and no less dangerous. The utmost vigilance was necessary to prevent the party from being separated or the boats lost. More than once in the darkness of the night, when anxiety would not let him sleep, Shackleton's sharp eye saw a crack before the watchman noticed it, and calling all hands he was able to see the whole company or the boats transferred to the same cake of ice. Once he says in *South* :

"The crack had cut through the site of my tent I stood at the edge of the new fracture, and, looking across the widening channel of water, could see the spot where, for many months, my head and shoulders had rested when I was in my sleeping bag. The depression formed by my body and legs was on our side of the crack. . . . How fragile and precarious had been our resting-place!"

Another time a crack ran right through a tent while the men were sleeping inside. Shackleton saw the tent stretch and tear as the fragments of the floe parted, and there was much difficulty in rescuing from the water two men who were engulfed while helpless in their sleeping bags Ten seconds after he had hauled the last man out of the water, the ice came together again like the snapping of a trap Nor was this the only rescue from imminent death effected by the Boss. By 9th April Patience Camp was near its end ; the cake of ice on which it stood had been reduced to a triangle, the sides of which measured only 90, 100, and 120 yards respectively. All arrangements for taking to the boats had been perfected; each boat had its allotted crew and leader, its share of stores and gear carefully planned beforehand, a sufficiency of preserved rations for forty days had been reserved for the emergency that had now arrived It must be remembered that the dwindling cake of ice was surrounded only by narrow lanes of water, often not wide enough for a boat to push through, often filled with a sludgy mass of half melted snow and ice fragments that would neither float a boat nor bear the weight of a man The killer whales, however, cruised in the lanes alert for prey This day a wider and clearer lane appeared, and the boats were launched and rowed towards the edge of the pack, which could be seen in the distance, and at night they were hauled up on another piece of ice where the tents were raised

The night was fraught with terrible experiences, so that it was a relief to get into the boats again in the morning, though some of the men were badly frost-bitten and all in a state of distress The three boats reached the open sea, but the waves ran so high that they were forced to get back into the lanes between the floating ice, again to camp upon a floe An effort was being made to work a way to the westward in the hope of reaching Deception Island now the sea was open

On the 12th, when Worsley succeeded in getting an observation of the sun, every one believed that in the three days the boats had made their way 30 miles at least to the westward of the point where they had deserted Patience Camp. Worsley worked out the position and Shackleton brought his boat alongside to receive the news. Worsley said nothing, but handed him the paper; the Boss looked at it and saw to his horror that the position was 30 miles to the eastward, and not to the westward of the starting-place; the drift had changed and they had been powerless to make way against it. He called out, "Not so much westing as we hoped for, boys!" and he very soon decided that the best chance left was to make straight for Elephant Island, 100 miles away.

On the 13th the pack suddenly opened and the boats were in the open ocean where a heavy sea was running. Shackleton in the *James Caird* commanded the flotilla, and Wild was with him, Worsley was in the *Dudley Docker*, and the first officer and Crean were the senior men on the *Stancomb-Wills*. They kept the three boats as near each other as it was safe to go, and all day battled with the rising sea. Some of the men were helpless with sea-sickness, so sudden was the change from the immobile floe to the wild tossing of the boats on the enormous waves. One or two were disabled with frost-bite, and a few were quaking with fear and had fairly lost their heads in delirium. So rapidly had the wind carried the boats steering north for Elephant Island away from the pack slowly drifting eastward, that there was not a bit of floating ice in sight, and they were short of water All night the boats lay one astern of the other riding to a sea-anchor, for Shackleton was afraid they might run past the island in the dark, and he knew that they could never beat back against the wind. There was a strong feeling amongst the men that they should go on and risk it, and Worsley thought it should be done; but he compelled obedience to the Boss, who for that night's inaction earned the strange nickname " Old Cautious."

That never-to-be-forgotten night was one of sheer horror. The spray deluged everything in the boats and froze in masses of ice which had to be chipped away in the darkness or the weight would have sunk them The temperature fell to zero Fahrenheit, which would try the staying power of a dry, well-fed, stoutly-clothed London policeman briskly pacing his beat on firm pavements ; here it fell upon enfeebled men wet to the skin, in worn and tattered clothing, crusted with ice, half starved and unable to shift their cramped positions for fear of upsetting the almost sinking boats. Shackleton did not expect that the weaker men could possibly survive the night They did survive the night and the next day too when the warm sunshine heartened them, though all were tortured by thirst, and at evening the cliffs of Elephant Island were only a few miles off. The *Stancomb-Wills* was in the worst plight, and the *James Caird* took her in tow. All that endless night Shackleton sat with his hand on the painter of the other boat so that he could know she had not broken loose, flashing signals to the *Dudley Docker* which had disappeared.

Next morning all three boats arrived together at a little strip of beach on the north coast of Elephant Island, and with almost incredible good-fortune every man landed safely and the boats were hauled up It was 15th April 1916; the last time any of these twenty-eight men had set foot on solid land was 5th December 1914, more than sixteen months before Here Shackleton after eight sleepless nights lay down on the beach and slept for many hours untroubled by damp or cold. On the 17th the whole party moved in the boats with much difficulty to Cape Wild, a beach which Wild had found on the previous day some miles to the westward The exhausted men were slow to take to the sea again, but Shackleton had found marks on the beach first reached which showed that spring tides completely submerged it Wild's beach ran up in a steep slope well above high-water mark ; at every other point along the coast the rollers broke against black cliffs which rose sheer for 1000 feet, or against the massive front of glaciers which filled every creek or valley.

Then followed three days of austere living and, for Shackleton, very hard thinking. No relief expedition would be likely to look for them on Elephant Island , it was hopeless to dream of an ocean voyage in the little open boats carrying all the men Two things might be done, either to send one boat with a few picked men to endeavour to reach some port, or else for all to remain where they were The second alternative had no hope in it, the first held just a glimmer, but enough for Shackleton to seize upon.

During the boat journey from the floe to Elephant Island his optimism alone upheld his followers. It was he alone who decided what were the reasonable risks that might be run and what the foolhardy projects that must be checked. He now decided that it was he who should take the path of greatest danger in the search for help, and that his companions should be the strongest men and best sailors, but that Wild must stay in charge of the party on Elephant Island as no one else had his power of extracting sustenance and a semblance of comfort from such conditions as existed there. Help must be sought at once before winter was on them and the island surrounded by impenetrable ice. The Falkland Islands were nearest ; but a boat could not lay a course athwart the westerly winds to reach them, and South Georgia, 800 miles away, could more likely be reached.

Shackleton decided to take the *James Caird* as the largest and most seaworthy boat. She was, in fact, only 22 feet long and about 7 feet at her widest, an ordinary whale-boat, very different in size and strength from the lifeboats carried on the upper deck of a liner. To improve her he had the services of a skilled carpenter in M'Neish, who strengthened the boat (the gunwale of which he had already raised in Ocean Camp) by fitting the mast of the *Stancomb-Wills* as a girder fore and aft to support a framework made of some old sledge-runners and box-lids, over which a covering of canvas was stretched and nailed down. Materials were scant and nails had to be drawn out of the venesta food-boxes in order to fix the canvas ; but in the end a creditable job was made of it and the fore-part of the *James Caird* was completely covered in. Ballast was put in, her mast set up, and provisions, the old sledging rations, for six men for a month, a cooker, some gallons of paraffin, and two casks of water. The crew chosen were Crean, M'Neish, M'Carthy, and Vincent, with Worsley as navigator. The arrangements were made with the utmost dispatch as winter was near—the penguins had departed to an easier climate, and the ice-pack was creeping up nearer and nearer to the island every day. Speaking of these experiences some months later in a letter to his son Ray, Shackleton said :

" It has been a much harder expedition than the last, and for months on end one never knew whether we would see the following day or whether it would finish up in the night."

All through he carried the weight of all the lives on his own magnificent optimism To any glimmering of insubordination he was stern, and the least docile member of the crew was cowed without a word by the mere glance of his indignant eye To weakness he was all gentle consideration, tending the men whose minds had given way under unbearable stress, like a mother. Every one depended on him and trusted him, and he entrusted them to the command and brotherly care of Frank Wild, who had never failed. On 24th April he started on his desperate venture.

The *James Caird*, with her company of six, hoisted sail and slipping through a lane in the on-coming pack reached the clear sea The twenty-two left behind sped the departure with three cheers and then set themselves to make the best of a very bad situation. In the months that followed there was never a wavering of confidence in the " Boss " or in his power of saving them. Their belief in his good luck went beyond reason, and they never thought that in undertaking the voyage to South Georgia he was deserting his comrades or trying to save himself So they remained lying or crouching under the upturned hulls of the *Dudley Docker* and the *Stancomb-Wills* while the *James Caird* went on her way.

Life on the *James Caird* can hardly be described, and it cannot even be imagined by those who have seen the huge waves of the Southern Ocean only from the deck of a liner. Those on board the little craft were already exhausted with the dreadful year of winter they had come through, their clothes were worn and tattered, their skin flayed at every joint with the horrible sea-blisters which salt water, cold, and the friction of rough cloth produce. They could not stand up, except for a moment or so, holding on to the mast or stays; they could not lie down except on the rough angles of the ballast and the cases under the dripping canvas "deck"; they could not even sit except in the open well at the stern, where the steersman on his two-hours' turn at the helm was often so cramped that he could not unbend his knees or lift his hands when relieved. Cooking was sometimes possible, one man holding the primus lamp, two squatting, one on each side, holding the cooking pot and lifting it clear when the worst lurches of the distracted boat threatened disaster. Whenever the sun showed at an hour when observations could be taken, Worsley took its altitude and worked out the position, managing in some manner peculiar to himself to set and clamp the sextant while he hung on to the mast with one arm and caught sight of the horizon from the top of a wave.

Shackleton remained alert and quiet but always cheery, seeking for any pretext for a joke as for hidden treasure, and splendidly seconded by Worsley's undampable spirits to uphold the men whose minds had not been trained to dominate the depressing drag of their racked and weary bodies. Down in the hollow of the waves the little boat would lie a while, shut into an illusive calm between two hills of water, from the summits of which the spume flew far overhead; a moment later she would rise on the crest and be flung forward by the shrieking wind in a smother of spray, rushing down into the next still hollow only to be hurled again into the tempest. The seabirds kept them company, little "Cape pigeons" which Shackleton could not shoot (he had his double-barrelled gun with him) because they looked so friendly. Great albatrosses, whose span of wing almost equalled the length of the little boat, swooped so low over it that the expression of impersonal interest in their hard, bright eyes could be seen, and it aroused a feeling akin to that of the Ancient Mariner after he fired the fatal shot, so they, too, passed immune. Strange sights had these great birds seen and grim struggles, but never anything stranger or grimmer than the *James Caird* and those she carried. And never could the spirits of his ancestors and

226 THE LIFE OF SIR ERNEST SHACKLETON

of his fellows, the old polar heroes, have looked with greater pride on Ernest Shackleton than now when midway on that supreme adventure. The labour of chipping off the ice as it formed gave place to ceaseless baling as the air grew less cold. When near despair in their worst moments, the men were cheered by the Boss's confident shout, " We're going to get through all right " But it was a very near thing.

On 8th May the fine navigation of Worsley brought the mountains of South Georgia into view, the little boat had been piloted across the great ocean to the very spot aimed at ; but a gale put her to a last rough test on the following day, and when the wind dropped it was found that the mast had lost its fastenings, and had the gale lasted another half-hour it would have fallen and probably ended the boat and this biography. The whole of 10th May was occupied by working back to the coast, and at dusk an opening between the cliffs appeared ; and as the water was quite exhausted and thirst was raging, Shackleton took a big risk and sent the boat in between the rocks and beached her in a cove. As he was making the boat's painter fast in the half-darkness, Shackleton slipped on the rocks and was badly shaken by a fall that might easily have proved fatal. Two of the men were in a very bad way, and all were absolutely exhausted.

When daylight returned it was discovered that the opening they had entered was King Haakon Fjord on the farther side of the island from the whaling stations. Nothing could be done for four days except to recruit the strength of the men by rest and food, and, fortunately, food of the pleasantest and most nourishing kind abounded in the shape of albatross chickens, not yet fully fledged, but large and plump and tender as turkeys. No thought of the Ancient Mariner now stayed their hand, though the sailors, vague as to the border-line between law and tradition, thought they were contravening some Wild Birds' Protection Act. To make a fire they broke up the " deck " and topsides of the *James Caird*, and so, at last, made themselves warm and nearly dry Then, on 15th May, they launched the boat and sailed up the inlet, landing close to the glacier at its head Here they hauled up the boat and

The *James Caird* approaching South Georgia, May 1916.

turned it upside down on a wall of turf and stones, creating Peggotty Camp

Shackleton found that two of the men were still too ill to move, and saw that it would be hopeless for the boat to face the rough and rock-strewn seas in the attempt to reach the whaling stations. So he determined to leave three men at Peggotty Camp and with Worsley and Crean to cross the unknown mountains in search of help The carpenter made a sledge to carry sleeping bags and food, but it was too heavy for the steep snow slopes, and Shackleton decided to make the journey which, from the chart, should not exceed 17 miles, in a single march. They started at three o'clock in the morning of 19th May, found a way up the snow-clad mountain slope, and after wanderings to and fro in search of passes, and up and down in darkness or in mist, they made their way across the Allardyce Range that no one had ever attempted to scale before, and on the forenoon of 20th May 1916 the three, shaggy, dirty, and ragged, reached Stromness Whaling Station. The feat was a miracle of mountaineering without guides or maps or resting-places The risks run were almost incalculably great, the toil enough to cloud their consciousness, and it is little wonder that more than one of the party felt as if they were accompanied by a Presence not of this world.

When they scrambled down the last cliff and staggered along the level to the settlement, the first inhabitants met, two boys, fled in terror at their approach; the next, an old man, looked at them and hurried away before they could speak. Then they came to the Manager's house, and Mr. Sorlle, an old acquaintance, stared at them in blank astonishment and demanded their names. "When was the war over?" asked Shackleton. "The war is not over," was the first news of the world of men after a year and a half of isolation. Mr. Sorlle proved a friend indeed. He gave them food, hot baths, and new clothes from the Store. In an hour or two they were civilized again.

If his return to the *Nimrod* on the Plateau, the Glacier, and the Barrier, seven years before, had been a race with Death on his pale horse, Shackleton's return from the *Endurance* over the Floe, the Ocean, and the Mountains, had been one long wrestling bout with the same grim adversary, dismounted, and in earnest. Never for an hour in all these months had Shackleton or his men been free from the menace which only unsleeping vigilance could save from becoming a strangle-hold. In this struggle Shackleton had risen to the height of moral greatness, though the ambition he had started with was wrecked and his party scattered. Battle by battle Shackleton had won so far; but his fight was not over.

The little boat which had played its part so stoutly was brought back to England, and for all time coming the *James Caird* will be an object-lesson in courage, resourcefulness, and unselfishness to the boys of Shackleton's old school, Dulwich College, where it lies, an inspiring memorial.

CHAPTER IV

ALL'S WELL. 1916-1917

"Such was this knight's undaunted constancy,
  No mischief weakens his resolvèd mind,
None fiercer to a stubborn enemy,
  But to the yielding none more sweetly kind,
  His shield an even-ballast ship embraves
  Which dances light while Nepture wildly raves.
His word was this : ' I fear but heaven ; nor winds nor waves.' "
PHINEAS FLETCHER.

THE bliss of returning to civilization, the warm bed, electric light, and solicitous kindness of Mr. Sorlle might have justified a few days spent in rest and recuperation ; but before he slept Shackleton saw a little steamer go off with Worsley on board to bring back the three men from King Haakon Fjord. Then he turned at once to the problem of relieving the twenty-two on Elephant Island. A steel-built whaling steamer, the *Southern Sky*, of 80 tons, was laid up for the winter at Husvik, but it was impossible to communicate with the owners in England. The magistrate, who represented the law and the government, sanctioned the use of this vessel on Shackleton's accepting full responsibility, and Captain Thom, whose ship was loading a cargo of whale-oil, undertook to command it, while a volunteer crew was soon made up from the whalers, all willing to stretch a point to save men whose plight might be their own any day.

While preparations for the voyage were going forward Shackleton heard from his host that the *Aurora*, while wintering in M'Murdo Sound, had been blown out to sea in May 1915, and drifted in the pack northward degree by degree simultaneously with the *Endurance*; but although seriously damaged, had escaped from the ice near the Balleny Islands in March 1916, and Stenhouse, the chief officer, had brought her safely to New Zealand. Ten men had been left behind, including Mackintosh and five companions who had not returned from their journey to the Beardmore Glacier.

Here was the complicated situation which now confronted him. Shackleton was on one side of South Georgia, three of his men on the other side of the island, twenty-two on Elephant Island on the point of being imprisoned in ice for the winter, ten men on M'Murdo Sound or lost for ever on the Barrier which had been fatal to Scott; one ship lost, the other seriously damaged. His funds, he knew, must by this time be exhausted, and he was isolated here with hundreds of miles of stormy sea between him and the nearest telegraph station from which he could let the world know his condition. Would the world care to know and bestir itself to help? The old world he knew certainly would, but the world was changed and had become strange. The calm Norwegians here in the whaling station told him of incredible things, of horrors unknown to civilization, of peaceful ships sunk by submarines, of towns far from the scene of hostilities ravaged by bombs from aircraft, of rationed food and censored letters, of compulsory military service and crushing taxation, of one country after another being dragged into the war or starved under the burden of neutrality "All the world is mad," was the summing up of the South Georgians, and it was to bring his people back to this mad world which was perhaps too preoccupied to help him that Shackleton had now to set his plans.

Already at home his anxious friends had roused the Government to action, and while he was recovering his strength in the cave on King Haakon Fjord a strong committee nominated by the Admiralty, with the Arctic veteran Sir Lewis Beaumont as chairman, had met to consider what steps could be taken to search for and relieve the survivors of the expedition. On the day Shackleton reached Stromness the committee reported, recommending that the old *Discovery* should be purchased and dispatched to the Weddell Sea, and the *Aurora* refitted for the Ross Sea. Representatives of the Royal Geographical Society,

Antarctic explorers, including Sir Douglas Mawson and Dr. W. S. Bruce, and his own representative, Mr. A. Hutchison, were giving their expert assistance to the committee ; but even had he known all this, Shackleton's sense of full personal responsibility would have forced him forward to exhaust his own energy before waiting for help.

On 22nd May, Worsley came back with the three men from Peggotty Camp, who in due time returned to England in the whale-oil ship, while on the 23rd Shackleton, Worsley, and Crean sailed in the *Southern Sky*. Shackleton was in high spirits. He had lost no time, and, if his luck held, the whole Weddell Sea party should be on their way home in a week. He told the story of the *James Caird* to hearers who knew the risks he had run, and could estimate the undaunted constancy of his resolution. He gloried in the achievement, for, as he put it, " I do not wish to belittle our success with the pride that apes humility." If the luck would only hold ! But on the 26th the ice-pack was encountered, and though he tried for two days to get round it there was no way. The little steamer would be crushed like an egg-shell if the pack had closed upon her. She could not be brought within 70 miles of the island, and as the coal-supply was running low, the attempt at rescue had to be abandoned, and a course was set for the Falklands On 31st May the *Southern Sky* entered Port Stanley ; Shackleton got his messages sent off by wireless, and at midnight, as Lady Shackleton was bidding farewell to two friends of the expedition at her house in Vicarage Gate, the telephone bell rang and she heard the news. Next day it claimed a larger share of public attention than one could have ventured to hope in that time of tense anxiety for war-news, only to be swept from most minds by the first alarming announcement of the Battle of Jutland before the authorities had recognized it as a victory. Nevertheless congratulatory telegrams poured in, foremost amongst them a warmly worded message from the King.

There was no suitable ship at the Falklands to undertake a winter voyage to the South Shetlands ; such a thing had never been attempted before ; but Shackleton knew that unless seals or penguins had appeared much later in the season than was to be expected, his men must now be on short rations and in great danger. The Governor of the Falklands, Mr. Douglas Young, was hospitable and kind, giving all possible assistance The Admiralty and several personal friends had endeavoured to find a wooden ship fit to enter the ice in some South American or South African port, but in vain. Many weeks must elapse before the *Discovery* could be sent out from England.

Sparing nothing on cables, which indeed he always used as carelessly as the old sixpenny telegrams at home, Shackleton got the promise of a trawler from the Government of Uruguay. Greatly to the credit of all concerned, this vessel, the *Instituto de Pesca No* 1, reached Port Stanley from Monte Video on 10th June, and Shackleton was away on her before night. The weather was bad, of course, nothing else was to be expected in those latitudes at that season. The Uruguayan officers suffered from the bitter cold, but did not turn back from the neighbourhood of Elephant Island, the peaks of which were sighted on the third day, 30 miles off, until every effort to find a way through the pack failed. At last, with the coal nearly exhausted the damaged trawler had to return baffled to Port Stanley The Uruguayan Government offered to have the vessel refitted and to try again ; but so much would have to be done that the delay practically meant abandoning the attempt at rescue. Nansen had offered, on behalf of the Norwegian Government, to lend the *Fram*, but time stood in the way of accepting this offer ; Peary was urging the United States Government to offer the use of the *Roosevelt*, but nothing came of it. Shackleton kept the wireless and cables lively, stirring up every seaport of the Argentine and Chile, asking the Argentine Government if their old gunboat, *Uruguay*, which had rescued Nordenskjöld twelve years before, was fit for the service, but she was not.

Then a mail steamer coming in, he crossed to the most southerly town of Chile, the little port of Punta Arenas in Magellan Strait Here is the centre of a considerable Scottish colony engaged in sheep-farming, and the British Association of Magellanes gave him a great reception, sympathizing with him to the extent of a subscription of £1500 towards a new attempt at rescue. The *Emma*, an old sailing schooner with auxiliary oil-engines, was chartered; a crew of ten men belonging to eight different nationalities (some hailing from places as remote as Finland, Andorra, and Mauritius), was brought together, and in view of the tempestuous weather the services of a little Chilean patrol steamer, the *Yelcho*, were secured to give a tow to the distance of 200 miles south of Cape Horn. It was hard work making way against the gales of the South Atlantic, and after three days of struggle in which the tow rope broke again and again, the *Yelcho* gave up off the port of San Sebastian, but caught up the schooner again at Staten Island and resumed the towing. A favourable wind enabled the *Emma* to make more speed than the *Yelcho*, so the tug returned to Punta Arenas, while Shackleton, with Worsley as navigator, held on his way in the *Emma*. By 21st July they had pushed on to within 100 miles of Elephant Island, where the ice enemy repulsed this third attempt as it had repulsed the two others, and after one of the most stormy cruises of their experience, the weary explorers struggled back to Port Stanley on 3rd August. Shackleton was pretty well worn out "I have grown old and tired," was the way he put it The news from home was not cheerful. His wife's sister Daisy, to whom he was devoted, had died on Easter Sunday, and her eldest brother, Mr C. H Dorman, who helped so notably in the *Nimrod* days, had died the year before; he felt the double loss the more as the news was not separated by a gap of time. To his daughter (aged 9) he wrote on 1st August :

"MY DARLING LITTLE CECILY,—Two years have gone by, child, since your old Daddy has seen you, and I am just longing to see my little girl again; though from your photo and from what Mummy tells me you have grown so much. It has been a long, long time to be away, and it has been a time full of work and danger, so that I want to come home and rest, and I want to walk with Mummy, you, Ray, and Edward in Kensington Gardens again and hear all you have been doing, and how you have been getting on at school, in fact, everything about you, darling. This is a funny little ship I am on now; we all live in a small cabin, and the water comes down through the roof or rather deck, as the ship leaks, and all the time she rolls about.

Just now I had to go up to alter the sails as a fair wind has come, and there is a chance that we may soon reach the Falkland Islands. We were not able to get through the ice to our men, and I have had to turn back, and now must wait for a bigger ship. I am very anxious about them, for they must have so little to eat now, unless they manage to get seals and penguins. We are very short of water and have not been able to wash since we left South America, three weeks ago; but that is nothing, for I had no wash from October last year until 20th May this year, and had not my clothes off from the 1st August 1915 till the 20th May 1916 I will have many stories to tell you of adventure in the ice when I return, but I cannot write them. I just hate writing letters, but I want you to get this to know I am thinking of you, my little daughter, and to tell you I loved your letter which you wrote last February to me. I know you will be a comfort and help to Mummy all the time I am away, and work well at school.—Good-bye, darling, thousands of kisses, FROM YOUR LOVING DADDY."

The news which awaited Shackleton at Port Stanley did not please him. The Government at home had taken action very promptly, as it seemed to those who knew the terrible load borne by the Admiralty during the war, and the *Discovery* was on the point of leaving Devonport for the Falklands under Lieut.-Commander Fairweather, and was to be expected about the middle of September. The conditions attached to the relief expedition seemed to Shackleton, in the absence of full information, to be unfair towards himself. The fact that the Admiralty Censor would read his private correspondence weighed on him and checked the freedom of his communications with his friends. A personal interview with the First Lord or with the Chairman of the Committee would probably have removed all misunderstanding, but that was impossible, and his mind remained troubled. By the end of September he was sure that the Elephant Island party would be in sore straits and some of them, he feared, would have succumbed to the hardship and hopelessness of their life. The Governor of the Falklands pressed the hospitality of his comfortable house upon Sir Ernest, but to him the prospect of walking to and fro along the bleak little port's one street, from the graveyard to the slaughter-house and from the slaughter-house to the graveyard, while his men were dying, was absolutely intolerable. They must be rescued, and he himself must rescue them before it was too late. He telegraphed to the Chilean Government begging that the *Yelcho* might be sent to take him and the *Emma* back to Punta Arenas, and the *Yelcho* came. The towing of the *Emma* across to the Strait in a north-westerly gale was a terrible experience, running very near disaster, but they reached the remote Chilean outpost in safety on the 14th and were met by the old kindness.

While waiting at Punta Arenas he wrote to his younger boy, then aged five:

"MY DARLING LITTLE EDWARD,—This is just a line from Daddy to give you his love and to say that I am just longing to see you again after all this long time Daddy will show you all the pictures of the ice and snow when he comes home again. Goodbye.—Your loving FATHER."

No suitable ship could be found at Punta Arenas, and after exhausting every possibility suggested by his quick imagination, goaded by fear for his men and resolve to rescue them himself, an improvement of weather led him to ask the Chilean Government to lend him the *Yelcho* for a last dash. He knew that the little steel-built vessel was incapable of threading the lanes of the ice-pack without risks that his caution would not let him undertake, and he promised not to allow her to touch ice if he could have her for his last throw for success. The Chilean Government had confidence in him, the Chilean people believed him capable of anything, and his heart was warmed by the readiness of their response to his call. The commander of the little boat, Captain Luis Pardo, cheerfully undertook the voyage, and with Worsley and Crean, Shackleton set out once more on his adventure on 25th August.

The comparatively favourable weather held and the *Yelcho* reached Elephant Island on 30th August, finding that the ice was open right up to the shore. Coasting along, Worsley sighted Wild's camp, and the imprisoned men sighted the ship

236 THE LIFE OF SIR ERNEST SHACKLETON

just as they were going to begin their midday meal. That meal was never eaten. A boat came ashore from the *Yelcho*,

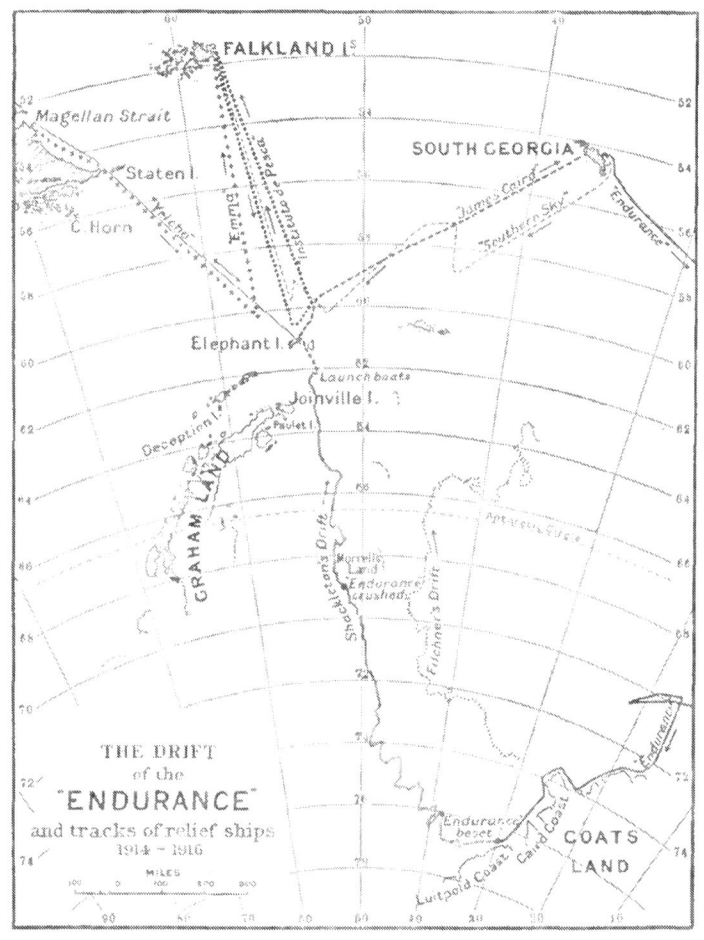

ROUTES OF THE *ENDURANCE*, THE *JAMES CAIRD*, AND THE RELIEF SHIPS OF 1916.

Shackleton in the bow, throwing packets of cigarettes to the men he knew must be in agony for tobacco.

"Are you all well?" roared Shackleton. "All safe; all well, Boss," shouted Wild, and even in their excitement the men on the beach noticed the smile that lit up his face as he breathed "Thank God!" As they had stood on the beach a few minutes before, gazing at the strange ship in some apprehension, they had recognized Shackleton's figure as he got into the boat, and the cry that rose from them was not one of rejoicing at their own release, but a fervent "Thank God, the Boss is safe!" This time Wild's "Pack up, boys. Perhaps the Boss will come to-day," which greeted each temporary clearing of the ice during that miserable winter, had been justified.

There followed a hurry of embarkation. Three times the boat went to and fro. The records of the expedition, the photographs, and the precious film from which so much was hoped, and all hands were soon on board, and one hour after reaching the camp the *Yelcho* was steaming north again, no man of the twenty-two in worse state and most of them much better than when he left them, four months before. They had had a hard time of it with storms and cold. Again and again they had seen their prison bars of ice break and float away, reviving hope, but as often the ice came back locking them in more securely than before, until this last time when the ship and the opportunity were on the spot together. Thanks to Wild's skill as a provider, to the constant attention of the two doctors, Macklin and McIlroy, and to the loyalty of the scientific staff and officers, there had been no loss of life and no breakdown of discipline. The spirits of the party had been kept up through many threatening hours by Hussey's cheery playing of that ideal instrument for desolate camps, the banjo. Shackleton's labours were rewarded; he grew young again in vindicating his leadership.

On 3rd September the *Yelcho* reached Punta Arenas, having announced her triumphant return by telephone from Rio Secco, a few hours before. The whole population turned out to receive them, the police having made a proclamation and rung the fire-bell to let every one know that great things were afoot. Every one felt a personal pride in the rescue, made in their ship, made by their people, and warm-hearted Chileans welcomed the strangers as they could never have expected to be welcomed even in their own country. They were conducted in a procession from the mole to the chief hotel by all the public officials of the province and a Chilean band playing the British National Anthem

Great men have passed by the sandy spit beside which the little town has grown up, and great incentives have fired their minds as they navigated the difficult channels of Tierra del Fuego, but to those people on that day, and to many the world over for all time, there was no unfitness in adding the name of Shackleton who saved his comrades to the roll which, beginning with Magellan, who discovered the Strait, and Drake in the *Golden Hind*, goes on to Charles Darwin of the *Beagle* and to Allen Gardiner, the ardent missionary. Never had the spirit of British pluck and pertinacity shone more brightly in the sight of a friendly people, proud of their own fine share in the gallant and glorious adventure

Once more Shackleton was at the end of a cable, glad to have saved his men from the Weddell Sea, scheming how to save those from the Ross Sea, terribly embarrassed by the exhaustion of the Expedition's funds, grievously troubled by the decision of the Australian Committee that he was not required on board the *Aurora*

As regards the Ross Sea Relief, the Committee was tripartite, responsible to three different governments which supplied the funds for repairing the *Aurora* and carrying out her voyage to the Ross Sea and back. The British Government paid half the cost, the Commonwealth Government of Australia and the Dominion Government of New Zealand shared the other half in proportion to the population of the respective countries. The New Zealand Government undertook the repairs of the *Aurora* at Port Chalmers and provisioned the ship. The Australian Committee had arranged that the Relief Expedition should be under the command of Captain J. K. Davis, than whom, in Shackleton's absence, no better leader could be found ; and on Shackleton's reappearance, though Davis, with the true instinct of a sailor, immediately placed his resignation in their hands, the Committee did not change its choice, and Davis continued to carry out the preparations. We believe that every one acted from a sense of duty ; and it is certain that Shackleton felt it to be his clamant duty as well as his legal right to command the relief expedition on his own ship, even though she had been refitted without his knowledge at the cost of others He felt that he must proceed to New Zealand with all speed and go to the rescue of Mackintosh and his men. The world was out of joint, and all the great sea-routes completely disorganized. He could not trust to catch a New Zealand boat in England in time, and so arranged to travel via San Francisco. As to the difficulty of finances, the Chilean and Uruguayan Governments had behaved with great generosity, and friends at home provided funds to pay off the members of the Weddell Sea party whose claims their Boss was determined should be the first to be met.

A pleasing duty was to thank the Presidents of Chile and Uruguay for their kindness, and the Chileans offered the *Yelcho* to take him and those of his party who proposed to travel that way to Valparaiso. On 27th September the little vessel entered the great bay round which the city rises tier on tier. Twenty-six years before he had seen that bay for the first time, when he was the least important person on the ship that brought him, the ship herself a mere incident of the day's business. Now the warships of the Chilean Navy were manned in his honour, and their flags dipped to the little *Yelcho* as she passed. Steam launches, electric launches, tugs, rowing-boats, all were out laden with a holiday crowd whose one desire was to see him, the hero of the hour, and they fell in line behind as his little steamer made for the quay, beyond which stretched the streets all crowded too.

Next day at Santiago the President of the Republic received him and presented him with the Chilean Order of Merit in the presence of 30,000 spectators. At night there was a State banquet, and the British Minister said to Sir Ernest, " Evening dress, of course." Shackleton protested that he had only the blue serge suit bought from the whalers in South Georgia ; but with his characteristic resourcefulness he borrowed the necessary garments from the members of the Jockey Club, all eager to have a hand in rigging him out. After the banquet a flash-light photograph was taken, and the guest of honour asked the President if he could have five copies. "Why do you want so many?" was the not unnatural reply "Well, your Excellency, you see I have on a syndicate suit, and I want every one who has a share in it to see the distinguished position his clothes occupied beside you" The President quaked with laughter at the quaint fancy, and each gentleman who had contributed an article of apparel got it back with a photograph bearing its name.

He gave two lectures in the capital, and, forgetful of his desperate need of funds, devoted the proceeds of one to the British Red Cross Society, and of the other to local charities. A special train was provided by the Government to carry him across the Andes, and so he entered the Argentine, and on to Uruguay where he thanked the President and Government at Montevideo Thence he went to Buenos Aires, where he was greeted and fêted as if, he put it rather touchingly, he "had made a triumph instead of having failed" in his expedition.

An important result of the wave of enthusiasm which Shackleton's knight-errantry aroused in the three southern republics was to turn the sympathy of many neutral and of some pro-German South Americans towards the Allies, and in this way it may well be that the ship's company of the *Endurance* served their country better in the Antarctic ice than they could have done in Flanders trenches

Early in October Shackleton was back in Valparaiso, and thence he sailed to Panama on the way by New Orleans and New York to San Francisco. On the steamer he wrote to his wife:

"I have had a tough time of it since leaving South Georgia. . . . I am old and tired, but you know I have been the means under Providence of carrying out the biggest saving of disaster that has ever been done in the Polar Regions, North or South."

More than once in his correspondence he exclaims, "I am sick of it all," and promptly adds, "but I must carry it through";

and again to a friend when on the voyage from San Francisco to Wellington:

"I am deadly tired of it all. I mean, of course, the worry that I have with the Committee, and cannot understand why they are so set against my trying to rescue my own men."

Outwardly, he was, as always, calm, cheery, and confident, determined to save his men in his own ship. Perhaps he did not realize that in his absence the Committees had been making all the necessary arrangements for the rescue of his men, and did not care to have their plans upset.

In Wellington, Shackleton was among friends, and very soon all New Zealand, the public, the Press, and the Government, declared its faith in him. His financial troubles were relieved by a few friends who got together £5000 as a personal loan, which Shackleton accepted with gratitude and reluctance, assuring the lenders that if he were killed in the war it could never be repaid. Dr. Robert McNab, Minister of Marine, who had never met him before, was greatly impressed by an interview with him, and did all he could to help. Shackleton was at first fiercely insistent on his rights, but when he saw that time was pressing he rose superior to his pride and his legal rights, and with a magnanimity amazing to those who knew him little, he accepted the conditions that had been laid down, which gave him charge of any land-journeys that might be necessary in the far South, but left Davis in command of the expedition. The New Zealand Government desired Shackleton to occupy the position of a passenger on board; but he felt that this might possibly give rise to trouble, and on the authority of Mr. L. O. H. Tripp, who was present, we are able to give in Shackleton's own words his reply to Dr. McNab:

"I am sorry if we differ on this point, but you must excuse me when I say I know more about the sea than you, and I know you cannot have two captains on a ship; and if I am on board and we get into trouble on the ice, half of the men will look to me, so I am determined that the officers and crew shall know that Davis is the captain and in entire command, and that I have signed on under him."

So he signed on as a supernumerary officer of the *Aurora* under Davis. He knew himself that this was as big a thing as he had ever done, one of the greatest triumphs of the spirit which animated his great-great-grandfather, the friend of Burke. He felt keenly having to leave Worsley and Stenhouse behind, but circumstances made this inevitable. Once on board the *Aurora* southward bound, his old happy temperament soon reasserted itself. He made things easy for his new captain and former subordinate, and Captain Davis assures us that, as of old, Shackleton was the most popular man on board the ship.

The *Aurora* left Port Chalmers on 20th December flying the blue ensign with the burgee of the Royal Clyde Yacht Club, evidence that legally she was Shackleton's yacht. She made a quick run to the Ross Sea, where on 10th January 1917 the old landmark, Mount Erebus, greeted Shackleton's eyes once more, with its calm ice-slopes and steady flag of steam. They reached Cape Royds with the *Nimrod* hut of happy memories, and Shackleton landed to visit the familiar scene. In the hut he found a letter stating that the Ross Sea Party had made their headquarters at Scott's wintering place at Cape Evans, opposite Inaccessible Island. A party of six men with a dog-sledge was seen approaching over the sea-ice from the south and soon they reached the ship. They had a moving tale to tell.

The southern party of 1915 had returned to Hut Point on 26th March, after laying depots in 79° and in 80° S. but could not reach Cape Evans as the sea was open, and the *Aurora* lay there frozen in, as all believed, for the winter. It was early in June before the ice was firm enough for the party to make the journey of 15 miles, and when they reached Cape Evans they found that the *Aurora* had broken away in a blizzard on 6th May before all the winter stores and equipment had been landed. Mackintosh, with nine companions, made the best they could of things for the winter, and early in October the depot-laying parties started the longest period of continuous sledge-travelling in polar annals. They laid depots in 81° and 82° S., and on 27th January 1916, Mackintosh, with Joyce, Ernest Wild, Richards, and Hayward, made their last depot near Mt. Hope at the end of the Beardmore Glacier to secure the safe return of the trans-Antarctic party, little dreaming that that section of the expedition had never landed and was now drifting on an ice-floe in the Weddell Sea, north of the Antarctic Circle, or that his own *Aurora* was fast in the ice north of the Balleny Islands, helpless to return to meet him. The return journey was more trying even than that of Captain Scott. Spencer Smith had been left helpless with scurvy in a camp several days before, and he was picked up and carried on a sledge. Mackintosh was weakening with the same disease. They were kept in camp for over a week with a howling blizzard Food and fuel were nearly exhausted when, on 23rd February, they made a desperate effort to march. Most of the men were too weak, so a camp was made for the invalids, and Joyce, Richards, and Hayward pushed on with the four surviving dogs, all very feeble. They just managed to reach the depot at the Bluff, 70 miles south of Hut Point. No party ever had a narrower escape from perishing, as their food was exhausted and they were hardly able to eat when the stores were reached. After resting they returned to their stricken comrades on 29th February in appalling weather. Mackintosh was now so weak that he, as well as Spencer Smith, had to be hauled on a sledge, and three days later Hayward was added to the load. The three men able to walk were very weak, and but for the dogs all would have perished. Thirty miles from Hut Point Mackintosh insisted on being left behind while the others went on with Hayward and Spencer Smith on the sledges. On 9th March Spencer Smith died after having been hauled by his comrades for six weeks over the Great Barrier; two days later they reached Hut Point. In a week the stronger men had gone back to Mackintosh's camp and had him in safety at Hut Point.

It was then 18th March, and there was no ship. Never was a finer record of work done according to plan, and a great disaster averted by the devotion of whole-hearted men. They had travelled more than 1500 miles in 160 days. The fresh food at the hut soon dissipated the scurvy symptoms, and in a couple of months the health of all five was restored. Mackintosh was very anxious to get to Cape Evans to find out what had happened to the ship, and who were left there ; but again and again just as the sea-ice was getting strong enough to travel over, a blizzard broke it up and drove it out. The party grew more and more disturbed at the state of things, and at last, on 8th May, Mackintosh and Hayward would wait no longer, but set off, the others not thinking it prudent to make the venture. A storm followed, and the tracks of the two men were traced to the broken edge of the ice, and then there was open water. The thoughtless daring which had carried Mackintosh through a life of dangers just escaped, had led at last to death. It was not until the middle of July 1916 that the three men could make their way from Hut Point to Cape Evans, where they joined the four who were wintering there, and now on the *Aurora* they told their story to their chief, whose heart was heavy within him to find that disaster had befallen this section of his expedition, though he was filled with pride, too, by the way in which the work they were sent to do had been done

Shackleton made several short excursions to search the coast and islands for any trace of the lost men, but none was found ; and on 19th January Davis took the *Aurora* out of M'Murdo Sound and made a rapid passage to Wellington, the port of return decided on by the New Zealand Government. Before the harbour was reached on 9th February, the world knew by wireless messages the tragedy of the Ross Sea party and the success of the last rescue Civic receptions greeted the returned explorers in the principal cities of New Zealand, and at the earliest moment most of them joined the armed forces of the Empire, eager to take their part in the war.

In accordance with the agreement entered into by the New Zealand Government, and concurred in by the governments of Australia and Great Britain, the *Aurora* was handed back into Shackleton's possession without any claim being made on him for the cost of her repair, outfitting, or maintenance. In the state of the shipping market a good purchaser was easily found, and before he left New Zealand Shackleton had the pleasure of repaying the generous and spontaneous loan which his friends in the Dominion had pressed on him on his arrival. This was a vast relief, but before Shackleton's mind was at rest he had to come to an understanding with the Australian Committee, and to express his sympathy with Captain Mackintosh's widow. The Committee met him in Melbourne; he demanded the reason of their exclusion of him from the command of the Relief Expedition, and they gave it. There was plain speech on both sides with perfect control of temper, and in the end handshaking and mutual respect. "I have buried the hatchet," cabled Shackleton to a New Zealand friend, and all that need be remembered of the episode now is its happy ending. Members of the Committee appeared with Shackleton on the platform when he gave his lectures in aid of Mrs Mackintosh, and the crowded audiences provided more than a thousand pounds, which Shackleton handed over to her.

Before leaving Wellington, Shackleton was persuaded to reduce to writing his personal experiences on the *Endurance* Expedition and the reliefs, lest they might be lost for ever if the explorer found the battlefields of Europe more fatal than those of the Antarctic. The pen was never a favoured implement with Shackleton, but happily Mr. Saunders, the congenial amanuensis who had taken down much of *The Heart of the Antarctic* from dictation, was available, and Mr. L O. H. Tripp has supplied this vivid and characteristic picture of how part of *South* was written ·

"I shall never forget the occasion I was sitting in a chair listening; Shackleton walked up and down the room smoking a cigarette, and I was absolutely amazed at his language. He very seldom hesitated, but every now and then he would tell Saunders to make a mark, because he had not got the right word; but that was only occasionally. I watched him, and his whole face seemed to swell, and I could see the man was suffering After about half an hour he turned to me and with tears in his eyes he said, 'Tripp, you don't know what I've been through, and I am going through it all again, and I can't do it.' I would say, 'But we must get it down.'

"He would go on for an hour and then all of a sudden would say, 'I can't do it—I must go and talk to the girls, or play tennis.' He walked out of the room as if he intended to go away, lit a cigarette, and then in about five minutes would come back and start again. The same thing happened after about another half hour or so. I could see that he was suffering, and when he came to his sensation of a fourth presence, when crossing | the mountains, he turned round to me and said, 'Tripp,' this is something I have not told you' As far as I can remember, his account of crossing South Georgia has practically not been altered in revision."

So the most thrilling chapter of *South* was written two years before the book could be published As always, the thought of home surged up, and in his last letter from New Zealand to his wife he wrote:

"I know you must have had a rotten time with many things, but I will do my best to see that all your troubles will be over and quietness in all ways be our portion. I have battled against great odds and extraordinary conditions for more than three years, and it is time that I should have a rest from it all. I would not alter or have changed one bit of the work and all its trials, for there is a feeling of power that I like, but at times I have grown very weary and lonely."

Shackleton had now completed his task, begun with such high hopes in 1913, complicated with such malignity of fortune by the outbreak of war in 1914, reduced to failure by the unprecedented absence of summer conditions in the Antarctic summer of 1914–15, spun out in the intolerable ice-drifts, the boat journey, and the repeated efforts to get back to Elephant Island in 1916, and now rounded off in 1917 by the return of the survivors of the Ross Sea Party.

Those who had criticized most severely the too ambitious scheme of the Trans-Antarctic Expedition were foremost in acclaiming the splendid generalship of the retreat, and the courage, endurance, optimism, chivalry, and self-repression of the leader. Nor was it forgotten that the expedition, though failing in its main intention, was not barren of results. It added to the knowledge of the difficult art of living in conditions bordering on the minimum at which life can be sustained. It tested new forms of equipment and rations for polar travel. It accumulated useful meteorological information, which has still to be fully worked up and brought into relation with existing knowledge , and it threw great light on the drift of the ice, both in the Weddell Sea and the Ross Sea. The depth of the Weddell Sea was ascertained at many points, adding to the information harvested by the *Scotia* and the *Deutschland*, and the mythical nature of Morrell's New South Greenland, which was clear enough to theoretical geographers before, was placed beyond the doubt of any reasonable person. Perhaps Shackleton himself would give the place of first importance to the strengthening of old friendships, and the formation of new ties by the expedition His confidence in his old and tried comrade, Frank Wild, required no confirmation, but Shackleton always gave to him the chief credit for the safety of the Elephant Island party. New friends like Worsley, Stenhouse, McIlroy, Macklin, Hussey, Wordie, and other members of the scientific staff who came into his life in the expedition, were destined to remain closely associated with him to the end; while the generous help extended by the people of South Georgia, South America, and New Zealand, and the friendships arising from it, brightened his later years.

Shackleton's brother-in-law, Professor Charles Sarolea, in a retrospect of his life and character in the *Contemporary Review*, says :

"In the popular estimate Shackleton was the spoilt child of fortune, the successful young adventurer whose face wore a perpetual smile because luck had always been smiling on him. And it is quite true that he did achieve an extraordinary measure of success. But we have hinted how adverse fortune had dogged his steps throughout his career, and how his greatness mainly appears in his cheerful and dauntless struggle against the obstacles which a cruel fate threw in his path. The one expedition which fully reveals him and which survives for all time in the annals of heroism, is the expedition which in a sense was a complete failure. ' Il y a des défaites triomphantes à l'envi des victoires.' "

CHAPTER V

[PRO PATRIA. 1917-1919

" To set the cause above renown,
　To love the game beyond the prize,
To honour, while you strike him down,
　The foe that comes with fearless eyes;
To count the life of battle good,
　And dear the land that gave you birth,
And dearer yet the brotherhood
　That binds the brave of all the earth."

H. NEWBOLT.

THE affairs of the ill-starred Trans-Antarctic Expedition being so far settled, all who had been on it were free to take their places in the fighting forces of the Empire, and Shackleton set out on his long journey home filled with patriotic ardour to place his services at the disposal of the Government. And as he went on his way he seized every opportunity to fan the flame of patriotism, whose gay spontaneous outburst had begun to waver under the threat of universal compulsory service. At Sydney, on 20th March 1917, he addressed a vast audience of 11,000 persons with tremendous effect, and the Australian Government printed a summary of the speech as a recruiting pamphlet, which was distributed by the hundred thousand, and produced far-reaching results. The speech was short and intense, the gist of it lay in these opening sentences:

" To you men and women of Australia I have something to say. I come from a land where there are no politics and no clashing of personal interest. For nearly two years I heard nothing and knew nothing of what was happening in the civilized lands Then I came back to a world darkened by desperate strife, and as people told me of what had happened during those two years I realized one great thing, and that was this:

" To take your part in this war is not a matter merely of patriotism, not a matter merely of duty or of expediency ; it is a matter of the saving of a man's soul and of a man's own opinion of himself.

" We lived long dark days in the South. The danger of the moment is a thing easy to meet, and the courage of the moment is in every man at some time But I want to say to you that we lived through slow dead days of toil, of struggle, dark striving and anxiety ; days that called not for heroism in the bright light of day, but simply for dogged persistent endeavour to do what the soul said was right. It is in that same spirit that we men of the British race have to face this war. What does the war mean to Australia ? Everything It means as much to you as though the enemy was actually beating down your gate. This summons to fight is a call imperative to the manhood within you."

Three weeks later Shackleton reached San Francisco, and found that the United States had entered the War on 6th April. During his short stay in America he addressed many meetings, striving to make the people at San Francisco, Portland, Seattle, Tacoma, Chicago, and Pittsburgh realize how closely their country was involved in the great issue being fought out in Europe, and helping to fan the rising war-spirit in Washington, Baltimore, Philadelphia, New York, and Boston. In America he was welcomed as an old friend, and the dinner given him by the Bohemian Club of San Francisco, with a series of complimentary songs written for the occasion, was one of the finest tributes to his popularity that he had ever received. His lecture in San Francisco drew an audience of 8000.

At New York he received the very rare distinction of being elected an Honorary Fellow of the American Museum of Natural History, a conspicuous tribute to his services to science

Before the end of May he was in London, happy to be home. He was received by the King, to whom he handed back the flag which, though it had not been carried across Antarctica, had flown over many forlorn camps in strange and perilous places, and had been shown with dramatic effect at recruiting meetings in the farthest outposts of the Empire.

Shackleton spent a week-end immediately after his return at the little house in Eastbourne, which his wife had taken as a measure of economy; but the sound of the distant guns in Flanders, which rose in the stillness of the evening and throbbed all night during those dreadful years, affected him so deeply that he could not stay, and rushed back to London to obey their call and enlist. He was persuaded to wait until a post could be found suited to his peculiar powers of organization and command, but he chafed at the delay He envied his comrades who were already in the thick of it; and to his nephew, Jack Sarolea, who had joined up as soon as his age permitted, he wrote:

" I have heard . of all your doings, and I send you this line to congratulate you on doing your bit in this great world-struggle. You have made good, and though it is stern and hard and means more than discomfort, you will always have the satisfaction that you are in the forefront, and every one there is the protector of the helpless ones who cannot fight. May you go on as you are doing, safeguarded by God, and win through, my dear chap.—Your affectionate uncle, ERNEST SHACKLETON."

It might have been expected that Shackleton would have taken to the sea again, but he had never forgiven the Admiralty for refusing him admission to the Navy in 1903, and he would not place himself at their mercy now. He had thought out a plan for a winter invasion of Germany and Hungary by a Russian army equipped for rapid movement over the snow by methods which he had perfected in the Antarctic, and for more than a month he remained in an alternation of hope and anxiety waiting for a decision on his plans from Petrograd. Nothing came of this scheme, nor of plans that had been put forward for his service in transport work on the French front, or in Italy, or in connection with the Ministry of Food in co-ordinating the supply of foodstuffs to the Allies; though this he considered " too soft a job."

Privately he was in distress as to the final adjustment of the financial responsibilities of his expedition, and the preparation and disposal of the kinematograph film, which was the chief asset remaining. It was hopeless at that time to attempt the publication of his book, and he refused to lecture except for charities, not wishing to make any profit for himself until the War was over. One lecture he did give at Ramsgate, where the benefactor of the expedition, Miss Stancomb-Wills, one of the town councillors, took the chair, and he was able to thank her publicly in her own town, where she kept up a hospital throughout the almost incessant air raids during the whole time of the war. He appeared at the Browning Settlement as its President, and he and Lady Shackleton were received by Queen Alexandra, whose interest in the expedition and the explorer was unfailing.

At length, about the beginning of September, the way was cleared for national service, but not on the War front. Sir Edward Carson, then the head of the Department of Information engaged in disseminating correct views as to the aims of the Allies in neutral countries, appointed Sir Ernest Shackleton, with the assent of the Foreign Office, to take charge of special inquiries in South and Central America. In a letter dated 22nd September 1917, Sir Edward defined the duties of this mission :

" (1) Thoroughly to examine the already existing propagandist agencies and to assist in spreading propaganda for the Allies.
" (2) To suggest to the Department of Information at the Foreign Office, after consultation with His Majesty's Ministers and Consuls and the local propaganda committees, changes or extensions which appear to you desirable."

These instructions were very wide, and left Shackleton with a free hand to carry out his investigations and formulate his suggestions in his own way. The wave of popularity which his efforts for the rescue of his comrades had raised in the three southern republics a year before had prepared the way for him, and assured him of a reception more cordial and a confidence in his personal qualities more complete than any other British subject at that time could command He liked the job, felt that he was specially fitted for it, and although the appointment carried no salary and required the whole of his time and attention, he welcomed it with all his heart. Strange work it seems for a man of action whose strength lay in struggling against physical dangers and hardships ; but the War demanded of most men new labour in unfamiliar fields. Nor was this altogether so unfamiliar as those who only knew Shackleton as an explorer might suppose ; for he had always been successful in personal negotiations with people of importance, and he knew his own powers.

The glad hurry of preparation was broken in on by a summons to Sandringham, where the King wished to hear the lecture on the *Endurance* Expedition, and see the photographs and kinematograph film. Everything passed off perfectly, and still glowing from the Royal visit Shackleton sailed from Liverpool on 17th October, on board the *Lapland*. The future looked more rosy than ever before, and his spirits were high. To the dangers of running the gauntlet of the German submarines he never gave a thought. Before sailing he wrote to his wife :

"Now . . . you are not to worry in any way about me. I will be all right, and will write to you at every opportunity, so it is not the same as going away last time in any way. . . . I think, darling, I will be able to speak to you once more on the telephone. I have the ball at my feet now, I must kick hard but carefully. All my love to you. Bless you. —Your own old ' Micky.' "

A final farewell from the steamer in the Mersey was full of gratitude to the friends who had helped him, and overflowing with love to his children whose letters gave him the most lively pleasure.

The voyage was long, as the route involved wide detours to avoid the danger of submarines, and Shackleton dismissed it briefly in a letter from New York as " dull, but safe." At New York he made inquiries which decided him to pass Central America by in the meantime, and proceed direct to Buenos Aires, as he believed the Argentine to be the country where British interests were most in want of attention at the time.

He sailed on board the *Vestris* on 3rd November with a crowd of American business men, all bent on pushing the trade of the United States in South America at this favourable juncture, when German trade was dead, and British trade strangled by war risks and restrictions. He landed at Barbados, Bahia, and Rio, where he felt the heat severely, and suffered from a bad cold and sore throat. Nevertheless, he lectured for a nautical charity in the saloon, and before the end of November he had reached Buenos Aires, and taken up his quarters at the Plaza Hotel. Here he was joined by Mr. Alan MacDonald, who had been so helpful to him in Punta Arenas the previous year, and whose knowledge of Argentine commercial life was now very useful.

The work was necessarily of an extremely confidential kind, involving the study of the personal character and foibles of the official and political personages in the capital, and the commercial policy and intrigues of the cosmopolitan mercantile houses. It was apt to touch the susceptibilities of the existing diplomatic and political agents, and certain to call forth the strongest hatred and opposition from all those engaged in German interests, the strength of which in South America was perhaps never fully realized at home. He found the methods of British propaganda to be slow, cumbrous, and terribly wasteful. In one store he saw nine hundred bales, weighing from 50 to 80 lb. each, full of printed matter, awaiting distribution, but all of it utterly out of date and useless for present needs. He insisted on having a telegraphic news service with local printing, and prompt publication of all matters of interest regarding Allied policy and successes. His first telegram from Buenos Aires to the Foreign Office asked for the immediate dispatch of five dozen model tanks, in order to let the wonders of the new arm become familiar, and teach the lesson of the superiority of the Allies in the invention of new and legitimate methods of warfare. Though now himself engaged in the exciting operations of a secret mission, he could not accustom himself to the institution of the Censorship, which seemed to paralyse his power of expression, even on purely personal matters in his private correspondence. Living in luxury in the great city he felt, he says, more homesick for his children than he ever did in the Antarctic snow deserts, and indeed it is strange to see how much his domestic affairs dwelt in his mind throughout all this South American sojourn. Not a letter home failed to contain kind messages to the old servants, one of whom had remained faithful to the family since they first set up house in Edinburgh, and in the midst of his exacting duties he was interesting himself in the welfare of her relatives

Early in 1918 the increased intensity of the British propaganda, largely due to Shackleton's work in connection with kinematograph films of the War, began to excite keen resentment from beyond the Andes, where German influence was paramount in Southern Chile. An anonymous letter from a German, daring Shackleton to come to Chile, where he would be promptly assassinated, was an invitation that he could not refuse. He went to Santiago on 20th January, and remained for a week in cool defiance of all threats No attack was made on him, and his quick response produced its impression. On another occasion his motor-car was only saved from a barbed wire that had been stretched across the road by the keen sight of the driver. As regards the two countries, Shackleton formed the view that the Argentines were more open to an appeal to their higher feelings with regard to declaring for the Allies, whereas, in Chile, commercial interests were preponderant, and the appeal which would be most effective there would be to the economic advantages of joining the winning side

On returning to Buenos Aires he resumed his activities in placing before the Spanish-speaking public a clear picture of German aims in South America, and the ambition to make the temperate portion of that continent a greater Germany. A map showing the political divisions of South America as they are, and as the German propagandists would like them to be, put this argument in a very telling form.

The summer of 1917–18 was very hot, and Shackleton was never happy in hot weather. He kept in condition by having three days of golf a week, starting on his round at 6 a.m. and returning for breakfast and the day's work. His final report was a document of great interest, but it is unnecessary to deal with it in detail, even if it were permissible to refer more precisely to a confidential paper. He made many suggestions for improving British prestige in South America, and retained to the end the good opinion of the British Minister, who acknowledged officially that he was most grateful for the help afforded to the Legation by Sir Ernest.

The work of the special mission to the South American republics came to an end in the middle of March, when Shackleton left Buenos Aires and travelled by Valparaiso and the west coast to Panama, closely attended by persons whom he suspected of being German spies, intent on observing and reporting his doings. When he reached New Orleans, the United States police arrested one of these followers and found that the suspicion was fully justified. On his way to New York, Shackleton stopped at Washington to confer with the British Ambassador, and towards the end of April he embarked for home, and refers thus to the voyage in a letter to a friend:

" At last I can sit down and write a line to you without the feeling that the Censor is at my elbow . . . I came from New York . . . in a convoy. We were twelve ships in all, and carried 25,000 United States troops. When we got to the danger zone, about 500 miles west of Ireland, we were met by seven destroyers; and it was a good job, because the next day we were attacked by two submarines, but before they could discharge a torpedo one of our destroyers dropped a depth charge and blew up one of the Huns; the other cleared off. We had 3000 troops on our ship. I was put in charge of No. 7 boat in case of emergencies, and given 50 soldiers for that boat, the colonel commanding the regiment, and the major, also the 8 civilian passengers, amongst whom were an Archbishop, his chaplain, an American parson, and four Labour Members I said to the captain, 'Why do you give me all that crowd?' He said, 'You know more about boats than any of us, and if anything happened, there would be a fuss if the Archbishop was lost, but there would be the dickens to pay if the Labour Members were drowned!' Anyhow, we got in safely, and I am glad."

He arrived home quite unexpectedly, as he would not let his wife know when he was crossing, to save her the suspense and anxiety; the early summer of 1918 being the most critical period of the war, with a maximum of submarine activity and daily news of the German advance in France on what was not yet recognized as their last great push.

During May and June Shackleton remained in London, writing reports and articles on South America, addressing meetings frequently in the country, including a stirring speech on the value of discipline to Boy Scouts on Empire Day, and with the authority of the Foreign Office fitting out a commercial expedition to Spitsbergen for the Northern Exploration Company He kept on offering his services for active war work to all the Departments, and at last was given a commission as Major. The first effect of this appointment was simply relief from the intolerable burden of appearing in the streets in civilian clothes while still in the prime of life and full activity. There had been irritating sneers in the frivolous Press at his long rest before joining up, the facts of the South American mission being necessarily kept from the public. One jibe was that as he had been seen about the Food Ministry on several occasions, he was hoping to be appointed Ice Controller.

The position for which he was destined by the War Office was to take charge of winter equipment for the North Russian Expedition, but between his appointment and the organization of the expedition he hoped to find time to accompany the Northern Exploration 'party to Spitsbergen. He left Aberdeen in the beginning of August on board the *Ella*, got across to Norway and through the Inner Lead among the islands as far as Tromsö, crossing the Arctic Circle for the first time, and strongly impressed by the contrast of open sea and warm summer weather with the ice-bound frigidity of his familiar Antarctic. Frank Wild had joined the party, and ultimately went to Spitsbergen as its leader. Another hero of the *Endurance*, McIlroy, who had been severely wounded in the war, had recovered sufficiently to go as surgeon, and Shackleton was full of delight at the prospect of being with them again in polar conditions. Delays for which he was in no way responsible had protracted the voyage and kept them long in ports where there was little

[Photo.] [Swaine.
MAJOR SIR ERNEST SHACKLETON, C.V.O., AGED 44.

Face p. 256.

to see and nothng to buy, for the neutrality of Norway had not saved that country from a food shortage far more severe than that at home. Just as the *Ella* was ready to set out from Tromso, Shackleton received an order from the War Office to return at once, as the equipment for the winter campaign of the North Russian Expeditionary Force had to be got ready with all speed.

He hastened back by the fastest route and, arriving in London at the end of August, he had six weeks of the most strenuous work of his life in getting the equipment together. The War Office adopted the system of shelter, clothing, food, and transport exactly as he had worked it out in his successive Antarctic expeditions. Tents, clothes, boots, sledges, cookers, and rations had to be ordered and their manufacture supervised, and on 8th October Shackleton was again at sea bound north, and dodging the submarines with as little thought of danger as when he was dodging the floes and icebergs of the Antarctic.

After some days Lady Shackleton received a telephone call from him at Newcastle saying that the ship would put into a Scottish port for a few hours, and asking her to come north for dinner with old friends. She set out by the first train and found the party assembled in the North British Hotel, when she reached Edinburgh It was one of the gayest leave-takings of that period of high-strung partings, and Shackleton, Worsley, and Hussey were boys together again, as they had been on the return from Elephant Island, while the friends of fourteen years before saw him as he was when first he alarmed the douce burgesses of Edinburgh by his impetuous unconventionality.

The field of action of the North Russian Expedition was on the Russian shore of the Arctic Sea along the Murman coast, which adjoins Norway in the Varanger Fjord, and is reached by rounding the North Cape, and in the White Sea lying farther east. A single-line railway runs from Petrograd northward to a terminus at Murmansk, near the Norwegian frontier, which is never blocked by ice. Archangel in the south of the White Sea is the terminus of another railway from the interior; but this port is inaccessible on account of ice, except for a few months in summer and autumn. There was a German army in Finland, the well-known objective of which was the occupation of the Murman region, with a view to establishing a submarine base either at Murmansk or Petchenga. There was also active opposition by Bolshevist forces, the Soviet Government having at length shown its hand, and declared against the Allies. The Allied Forces, consisting at first of a mere handful of troops and of naval ratings landed from the warships, had been reinforced by various Allied contingents and strengthened by the enrolment of local recruits, but they were still woefully weak for the task assigned them. Those at Archangel were under the command of Major-General W. E. Ironside, whilst those at Murmansk were commanded by Major-General C. M. Maynard, to whose staff Major Sir Ernest Shackleton was attached as Advisor on such matters as equipment, clothing, rations, and transport for mobile columns. He was given a free hand in his work, and to his great delight had secured as assistants a number of his old polar comrades. Worsley and Stenhouse were lent by the Navy, and Hussey and Macklin transferred from other parts of the Army. " Syren " was a code word indicating the Murmansk force, as " Elope " indicated that at Archangel.

The journey north on the transport was slow and cautious, but it was like going home to meet the sharp breezes of the Arctic after the harassing year with its two exhausting summers, the first in Buenos Aires, the second following on in London, with no winter between.

As he reached the polar regions again with all his new duties pressing on him, his thoughts were of home, and he wrote to his son Ray, aged 13, who was preparing for his first term at Harrow:

" We are now up in cold weather, but it is clear and fine, and at night there is a wonderful aurora swinging across the sky. The moon lies low on the horizon, and circles the head of the world. You could write a poem of all the glory of this wonderful North. I want you to send me copies of any that you have written; and do keep it up, for the love of poetry is good, no matter whether one's life is carried on at home or in the wild places of the world. I am sure you will like Harrow; and even if the comfort is not much you will get used to it all, and to be amongst boys who some time may be great in the work of the world is good. You know, my dear boy, there is no reason why you yourself should not get on also; work now, that is the main thing; but keep up your boxing as well, for that combines all exercise. Mummy wants you to be good at cricket too, and at Harrow you will have opportunity enough. . . . Write to me occasionally, and when home for the holidays keep things nice and smooth, for you are the eldest, and you are the man in the house, as I am so much away. God bless you, Sonny.—Your loving DADDY."

Shackleton tried to settle down at General Maynard's headquarters at Murmansk, telling his friends he had at last got a job after his own heart—" winter sledging, with a fight at the end of it "; but sad to say no opportunity for armed encounters came his way, and the man who all his life had been a fighter by nature, was in the end denied what would have been the crowning happiness of meeting his country's enemies in the field. As things turned out, it proved imperative that he should remain at headquarters occupied with rather humdrum work on stores and instruction in the use of Arctic equipment, but he found the quarters pleasant—" more of a happy family than a rigid mess," and the country round was novel and attractive in the polar night, with its snow-covered hills clothed with dwarf forest, and beautiful lakes in the hollows. So his friends pictured him as winter drew on, watching over the mobile columns which were to keep the lawless countryside in order, now that the Armistice of 11th November had stopped the Great War; but one night in December, as Lady Shackleton was crossing the dark hall of the house in Eastbourne, the door opened and he was there.

General Maynard, who was in command at Murmansk, had been summoned home, and Shackleton had accompanied him on a fast cruiser, one of their tasks being to get fresh equipment to meet the new conditions, for the force was not to be withdrawn at the end of the German menace. The short stay in London was a tremendous rush of official work, with snatches of the wild excitement that submerged Society when the heavy restraints of four years of war were suddenly relieved Shackleton was in close attendance on his general, giving him the loyal and affectionate service to which he had long been accustomed from his own tried followers, and accompanying him to Buckingham Palace for an audience of the King before returning to the far North.

The way back was by rail to Invergordon, for the Cromarty Firth was still an important naval base, known merely as "somewhere in Scotland" On Christmas Eve, Lieutenant A. S. Griffith, R N R., of H.M.S. *Mars*, received orders to land a working party to transfer the baggage of General Maynard and Major Sir Ernest Shackleton from the naval train to the transport *Umtali*, which was lying in the firth It came to him like a flash that this was the Mike with whom he had played truant from Dulwich College in the Sydenham woods more than thirty years before, with whom he had plotted to run away to sea, and whom he had never met again The meeting of Mike and Griff can be imagined. They were both boys again, and their work finished, they spent Christmas Day in living over the old times when they hatched plans of wild adventure which, marvellously indeed, had fallen infinitely short of reality. Shackleton noticed a telephone on Griffith's table in the *Mars*, and said he would give a lot for a last talk with his wife. Arrangements were made for the line to be cleared through to Eastbourne for ten minutes at 1.30 p m , and prompt to time the conversation took place through 650 miles of wire, and every word distinct. "Never write when you can phone," was always his maxim Lady Shackleton was just beginning to carve the Christmas turkey for the expectant children, when the telephone bell broke in on the feast.

That night Shackleton joined in a regular sailors' Christmas party in the ward-room of H M S *Dublin*, and never had he been more boisterously exuberant than amongst his new naval friends, with his old chum by his side.

On his return to Murmansk he resumed the routine duties of his post, and was satisfied to find that his equipment had practically stopped frost-bite. The men with whom he came in contact responded to his sympathetic appeal. One of them, writing long afterwards to a newspaper, said:

" How I recall his striking figure during the North Russian campaign daily exhorting by his magnetic influence suffering humanity to greater tasks. . . Eccentric in some ways; almost totally unheedful of cold, and clothed lightly for such parts, Shackleton forced upon all whom he encountered a lasting impression of real merit. An idol of the mobile columns, an inspiration to all, he aided materially the *moral* of the troops and effectively equipped the entire Russian Force against the rigours of winter with a scrupulous thoroughness "

How he impressed his superiors may be judged from Sir Charles Maynard's recollections, which cannot be presented better than in his own words :

" I must admit that I heard the news of Shackleton's appointment with somewhat mixed feelings Whilst I was only too anxious to get the assistance of some one with a wide experience of such conditions as those obtaining in North Russia during winter, I was not certain that Shackleton was the right man for the job. I had never met him, but the impression I had gathered from hearsay was that he was somewhat dictatorial if not overbearing, and that, though doubtless a fine leader of men, he was unlikely to accept gladly a subordinate position. Events soon proved, however, that my fears on this score were totally unfounded, for from the moment of his arrival to the time of his departure in the spring of 1919, he gave me of his very best, and his loyalty from start to finish was absolute. He fitted at once into the niche awaiting him, and both he and his friends of past Antarctic expeditions who were working with him laid themselves out unreservedly to further the interests of my Force

"As was natural, Shackleton had a great hold on the imagination of the men, who were keenly interested in hearing him tell of his past experiences, and ready to take to heart the lessons he inculcated. He proved a cheerful and amusing companion, and during the long, dark winter of 1918-19, his presence did much to keep us free from gloom and depression. He became almost at once a member of the happy family of which my Headquarters Staff consisted, taking and giving his full share of chaff, and pulling his weight in helping to provide amusements and entertainments for all ranks.

"Needless to say, I found his expert knowledge of the utmost value, and the equipment provided on his advice proved entirely satisfactory. His experience was especially useful in drawing up regulations for the loading and packing of sledges for employment with mobile columns.

"My private relations with him were of the most cordial nature throughout the whole of the time in North Russia. I admired him for his splendid record as a leader, and as a man of strong will and determination; but I also formed a very close attachment with him as a personal friend. It will be long before I forget the hurried trip to England which I paid with him in December 1918. Both on the journey and during our stay in England he insisted on acting as a sort of glorified A D C to me, doing all in his power to help me in official matters, and to save me worry and trouble over private arrangements Here was a very different Shackleton from the Shackleton I envisaged when I was first informed that he was to be attached to my headquarters

"Whilst with me, Shackleton showed himself a man of extraordinary energy and of many interests. Quite apart from his work of exploration, he took an active part in many schemes, having for their object the development abroad of British trade, and though doubtless—as was only natural—his own business interests were those which concerned him chiefly, he always had at heart the interests of the Empire as a whole. For the British Empire was to him a very real thing, a heritage of great worth, the maintenance and expansion of which was the primary duty of every Britisher."

Early in 1919 Shackleton had satisfied himself that the North Russian Force would not be likely to pass a second winter in the far north, and that his usefulness in that field had come to an end. He felt therefore that he was justified in resigning his commission, which he did on 9th February. He had seen something of the potential wealth of Northern Russia, and felt that it would be a great thing for British trade if a concession could be obtained for the development of a large tract of land by British capital At the same time he recognized that if he could negotiate such a concession from the existing government of North Russia, which remained independent of the Moscow Soviet, he would at one stroke solve all his own financial problems, and find the fortune that he always believed he was destined to enjoy Accordingly, being free of the army, he made a remarkable overland journey from Soroka on the Murmansk railway to Archangel, a distance of 200 miles through forests and across great snowy wastes, and there he succeeded in his scheme, and saw the future golden before him as he had seen it so often before. Had not the Moscow Soviet proved far more powerful, and the Northern Soviet weaker than could be suspected at the time, both he and his country, and above all the people of the North, would have profited immensely. The disappointment, like the Nemesis of all his happy dreams, came later. For the moment there were difficulties enough to make life keenly worth living, and hopes enough to compensate for all rebuffs. At Archangel he met his old friend of the *Nimrod* expedition, Dr. Eric Marshall, who, ever since his return from the Antarctic, had been exploring in the remotest forests of the tropics until the War called him home Then there was the return journey to Murmansk in the polar dawn at the end of February, and a month later, Shackleton was home again after one of the stormiest voyages of his experience in crossing the North Sea.

The War was over and Shackleton was free to turn to his own affairs, which had been much neglected and were sadly embarrassed. In April he was able to spend four days with his family at Eastbourne, " the longest time at home for the last five years," he declared in a letter. He was suffering from a bad attack of neuritis and was very tired. The prospects before him were dark, but his indomitable optimism shone as cheerily as ever. He set himself to lecturing all over the country " to keep the pot boiling " ; he gave much time to the correction of the proofs of *South*, which was at last published in November. He wasted weeks of superhuman labour on the North Russian Concession, which had been so full of promise in spring, but withered away with the fall of the leaf. He took up a new and most attractive scheme for the cheap production of agricultural fertilizers on a large scale.

In autumn, during the railway strike, he was remobilized for the purpose of organizing the London omnibus service in case there was a general strike of transport workers ; but happily this did not arise. He suffered frequently from chills and sciatica ; but only mentioned this to sigh for the health and freedom from worry in his beloved Antarctic The scheming and selfishness of the business world were peculiarly hateful to him, and indeed he was singularly unfortunate. *South* proved an immediate success so far as sales went ; but he had been forced to assign the royalties to the executors of one of the guarantors of the expenses of the *Aurora*, who had verbally assured him that if disaster overtook the expedition his loan would be viewed as a gift. But he had died, his sons had been killed in the War, and the executors took a strictly legal view of their rights The effort to get a hall for the exhibition of his film taxed his strength and temper sorely, and no use could be made of the greatest asset of the *Endurance* Expedition until the year was almost done Even then the superior business acumen of others promised to make his share of the proceeds of the work which he had risked his life and pledged his fortune to produce, a mere fraction of the prospective profits In the midst of the luxury and excitement of London, Shackleton was harder put to it than when drifting on the floes in the Weddell Sea, or crossing the mountains of South Georgia ; yet through it all he kept to the faith that he was only " baffled to fight better "

CHAPTER VI

THE LAST *QUEST*. 1920–1922

> " We are the fools who could not rest
>   In the dull earth we left behind,
> But burned with passion for the *South*
>   And drank strange frenzy from its wind
> The world where wise men live at ease
>   Fades from our unregretful eyes,
> And blind across uncharted seas
>   We stagger on our enterprise"
>   ST. JOHN LUCAS as quoted by E H S

THE first exhibition of the film illustrating the voyage of the *Endurance*, and showing the long-drawn-out agony of her final destruction and engulfment by the ice, was given in the Albert Hall on 19th December 1919, when the Earl of Athlone presided over a large audience. The boat which carried the forlorn hope from Elephant Island to South Georgia was to have been on view in the hall, and it had been brought to the doors on a lorry where it had to remain outside, because it could not pass through the doorways. The *James Caird* arrived at the last moment, or Shackleton would have had it in the destined place on the platform; he had often done more impossible things than to take a boat to pieces, or widen a doorway, but to do so requires a little time The lecture was a success; never before had an audience seen such moving scenes under the guidance of the man who had lived through them. By a flash of his true character which baffled the comprehension of those who failed to see his greatness, the whole proceeds were given to the funds of the Middlesex Hospital

One of the few short rests of his life included a Christmas at home at Eastbourne, where Shackleton appeared in the congenial rôle of Father Christmas for the special delectation of his youngest boy, who sleepily refused to look at the stranger trying to rouse him in the middle of the night. The rest was short, and before the end of the year Shackleton had embarked on a series of lectures on the *Endurance* Expedition, illustrated by his marvellous film, at the Philharmonic Hall in Great Portland Street. He carried on this work almost without a break for five months, giving two demonstrations of two hours each daily for six days every week Towards the end he added a fifth hour per day, to give special shows to the children of the London County Council schools. The physical exertion was enormous, and he felt it tell on his strength ; the mental strain was also great, for he practically lived all those months with the wreck of his ship and the shattering of his hopes always before his eyes Repetition, of course, brought a certain merciful callousness, but the undercurrent of a sense of failure so repugnant to his nature could not be stopped The few occasions on which he was compelled to ask Wild to lecture for him, a duty which that staunch friend never shirked, however little he liked it, brought small relief One of these involved a journey to Scotland to give his lecture to his old Society, the Royal Scottish Geographical, at Edinburgh, Glasgow, Dundee, and Aberdeen Several times he was summoned to the deathbed of his father, who had been hopelessly ill for two years ; but twice he got away to Harrow-on-the-Hill to see his son Ray, and address the boys ; and once he stole one of his precious Sundays to greet the Browning Settlement at Walworth of which he was still the president

The hard labour of the platform was not without its brighter touches. The hundredth lecture was made the occasion of a reunion of his old comrades, no fewer than nine of them supporting him on the platform, and after the lecture " the Boss and the boys " had a great time together, the Elephant Islanders reproducing one of their musical evenings, with Hussey on the banjo once again, as so often in the dark days, the sole instrumentalist. One day in March the Philharmonic Hall was filled by Fellows of the Royal Geographical Society, when Sir Francis Younghusband, the president, took the chair and assured Shackleton of the esteem in which he was held by the

[Photo.] [Vandyk Ltd.

Society. And every day questions were invited from the public, some of them weak and silly enough, with now and again a tough heckler who recalled the bygone thrust and parry of the Dundee election, and stung the lecturer to speak his mind. One of these raised a question which got an answer that deserves to be put on record.

A man in the gallery called out, "What are your men doing now?"

"Different jobs."

"And while they are working hard you are making money here out of what they did for you!"

To this Shackleton's reply was—"You have raised a delicate question, and one I do not usually discuss, but I will answer you. You may not be aware that when a ship founders all pay ceases automatically; that is the law. But not only were my men kept on full pay for a year longer, for the sake of their dependants, but for three months more after their return. It is to pay those liabilities that I am giving these lectures, and you will be glad to hear that the shilling you paid for your seat (if you did pay it) goes to that object"

The lectures were greatly appreciated by the public, and curiously enough the audiences were mostly men, but the attendances were irregular, and the takings sometimes failed to meet the expenses for weeks together. Shackleton was sorely in need of money, as in addition to the expedition debts and his own expenses he had heavy demands upon him on account of his father's long illness. When the lecturing period was nearly over he wrote to a friend, "It is a strain, but then all my life is a strain, and I would not have it otherwise," and indeed he enjoyed the fight of it

Quite early in the year, as soon as the initial difficulties at the Philharmonic Hall were overcome and the lectures going automatically, Shackleton came to the decision that he had strength enough left and life enough before him to carry through yet another expedition. Every time that he was left in the lurch by the clever men of business with whom he tried to compete, his mind went back to the one sort of enterprise in which he was supreme. The fact that "the boys" who had waited for him on Elephant Island and rallied round him from the battlefields and the mystery ships were ready to follow him at a word, seemed of itself an invitation to go out again. At the beginning of March he was speaking confidentially of two alternative schemes One was to turn to the other pole and explore the unknown area of the Beaufort Sea lying to the west and north of Prince Patrick Island. The other was to set out on an oceanographical expedition with the object of visiting all the little-known islands of the South Atlantic and the South Pacific, and searching again for the reported islands which at one time or another had figured on the charts, but which no living sailor had ever seen. This was an old idea, for when he sent the *Nimrod* home from New Zealand under Davis in 1909, it was with instructions to hunt for all such doubtful islands as lay near her track, and the work had been done diligently though with negative results.

"The boys" were eagerly hopeful, and although Wild and McIlroy started to prospect in Central Africa with a view to ultimate settlement there, and Worsley and Stenhouse were away on a trading venture of their own, they were all "at call," ready to join in any expedition at short notice

The bondage at the Philharmonic Hall ceased at the end of May, after two hundred and fifty appearances, and the summer was devoted mainly to the hunt for a ship, and for a benefactor to provide the necessary funds. It had taken a long time before Shackleton was "battered by the shocks of doom" into distrusting an unsecured future. But this time he resolved that he would not take an irretraceable step until he had money secured as an absolute gift for the purpose of an expedition. With revived keenness he sought for introductions to the richest men in the country, an easy matter now with the reputation he had achieved; but the men of millions heard his glowing plans unmoved, they were repelled alike by the North Pole and the South; to them the lonely islands of the ocean made no appeal, unless the prospects of new wealth lay there Shackleton could not promise anything but glory as a reward; the rich men knew they could get all the glory they wanted at a cheaper rate

Then on an eventful day two Dulwich College old boys met after long separation, and Mike, still a boy at heart, woke an answering chord in John Q Rowett who had prospered in the world, but still held his old friend in warm affection. Mr. Rowett agreed readily to contribute a very handsome sum to a new expedition, provided that others would do their part.

At this stage Dr Shackleton died, leaving a happy memory behind him in the hearts of a multitude of old patients and friends, but nothing more substantial. Sir Ernest took all the burdens upon himself, heedless of his own embarrassments; but he did not abandon the expedition which he knew well must be his last, however successful it might prove.

The Beaufort Sea plan was now the more attractive. Shackleton had met the Canadian explorer Stefansson in the spring, and discussed the conditions of the problem with the man whose courage, resourcefulness, and originality had promised to revolutionize Arctic exploration. Now he had got into touch with the High Commissioner of Canada, who favoured the idea of a northern expedition. So during the autumn, Shackleton was travelling all over the British Isles searching for a suitable ship in every port, and over a large part of Europe to track down potential patrons of discovery in their holiday resorts. The result was almost always disappointment, but disappointment never hardened into bitterness, it evaporated without leaving a stain under the warmth of a new expectation On his flying visits to his home, his boyishness often made him seem no older than his children. One day he went with them and Lady Shackleton, who had thrown herself heart and soul into the Girl Guide movement, to a camp at Eastbourne. A large lunch basket accompanied the party, and when it was opened the *pièce de résistance* was found to be a big toy penguin, which the explorer had carefully packed, with a carving-knife and fork, to ensure hilarity at the meal.

At length, on a visit to Norway, he found a ship well suited for an Arctic expedition, a little Norwegian sealer named the *Foca I* only 111 feet long, with 24 feet beam and 12 feet depth of hold, schooner rigged, with auxiliary engines and a tonnage considerably less than 200. She was purchased before the end of the year, and renamed the *Quest*.

The year ended with one of his rare Christmases at home. Thanks to the generosity of Mr. Rowett and of Mr Frederick Becker, who had given £5000 to the funds, the expedition was almost assured, though no public announcement was made

On 18th January 1921 he wrote as follows to an old friend :

" It was good to see your writing. I always feel that through the long years you have been one of my strong supporters, and can never forget that I made my first dip into the sea of science under your tuition. In those days ' the world was a steed for my rein,' but as we grow older we grow more humble, and now from the stern Antarctic and years of isolation, I am turning to the milder North (for your ear only at present). I am about to launch my last polar venture, to wit, the Exploration of the Beaufort Sea I am just off to Canada, but will be back on the 15th of February, and will then tell you everything "

The visit to Canada resulted in promises of support which appeared to justify the equipment of the expedition, but nothing was published, and when the plans were well advanced and the old *Endurance* comrades brought together, the news from Canada got less satisfactory, and Shackleton had to cross the Atlantic again This time it became too clear that the promised Canadian support was to fail, and that the summer would be lost for exploration. It was a terrible disappointment, but it showed up the unaltered character of the man. One plan having failed, he fell back on another without an hour spent in useless regrets Nothing had been published except vague rumours of the *Quest* being fitted out for an expedition to the North, and Shackleton had neither confirmed nor denied the truth of these when newspaper men assailed him. He followed up the letter just quoted by another on 24th June :

" Here follows a surprise for you. . . .
" It became too late for me to go North this year, and on Wednesday morning next you will see the announcement of an expedition after your own heart—an oceanographical and sub-Antarctic expedition, fully equipped, fully financed, seventeen officers, no A B.'s, all men keen on work,

"Programme in a nutshell : all the oceanic and sub-Antarctic islands. Two thousand miles of Antarctic outline from Enderby Land to Coats Land Seaplane, kinema, wireless, everything up to date, but still wanting your advice and expert knowledge. " I have worked swiftly and silently."

Shackleton continued to work swiftly, for the time for preparation was very short, but the silence was broken by a burst of publicity for the expedition and its plans which attracted the widest interest Incidentally the insistence in the daily Press on the great results which were to be expected from the voyage of 30,000 miles planned out for the *Quest* abated the sympathy with the expedition of some critical scientific folk, and the whole thing was sometimes referred to contemptuously as " an advertising stunt." This was quite unjust. Shackleton certainly did not shun the public eye, and he heartily enjoyed the bustle and excitement of preparation, but he was always determined that his work should be fully deserving of what praise it earned, and his plans equal to the expectations they aroused. It was necessary for a time also to continue the search for funds. Eventually Mr. Rowett added to the generosity of his early contribution by assuming responsibility for the whole remaining financial obligations of the enterprise. So the Shackleton - Rowett Expedition came into being. The history of recent British polar research has been rich in generous patrons, Lord Northcliffe, Sir George Newnes, Mr. Ll. W. Longstaff, and Sir James Caird have all honoured themselves by large and unconditional gifts ; but none of them gave more largely or with more devoted a trust in the expedition's leader than Mr. J. Q. Rowett, whose generosity swelled to a splendid total. Shackleton was now able to complete his preparations with a mind relieved of long-standing financial worry This set free an effervescence of boyish exuberance which delighted the hearts of his friends.

Time was ever his enemy. To ensure the success of his fine plan of exploring the Enderby Quadrant of the Antarctic where no mechanically propelled vessel had ever tried to penetrate the pack, it was essential that he should be on the Antarctic Circle, 1500 miles east of the Cape of Good Hope, by Christmas

Day To ensure the fullest advantage from the proposed island surveys and oceanographical observations it was necessary to collect apparatus and train observers, which would involve months of preparation and postpone the expedition to the following year But the ship was there, the men were waiting impatiently, and perhaps we may infer that something within him warned Shackleton that he himself might find next year too late In the summer of 1921 he was confident in his physical fitness, the occasional trouble from colds, from indigestion, from sciatica, from rheumatism would vanish he was sure as soon as he got to sea again with a ship of his own and " the boys," but next year perhaps——. It was now four years since in his confidential letters he had confessed that he would be too old for another Antarctic adventure ; and it may well be that he recognized, though he did not acknowledge, that this was his last chance to do great things in the field where alone he felt sure of success.

So the preparations were pushed on with fevered haste. Never was a little vessel crowded with such a wealth of the latest appliances. Wireless installations of the most ambitious range, gyroscopic compass, electrically heated overalls for the look-out man in the crow's nest, a seaplane, three large boats and several small ones adapted for landing on surf-beaten shores or conversion into a sledge on ice-fields, kites for the meteorological exploration of the upper air, sounding machines and gear for collecting specimens of everything collectible in air or sea or land. The *Challenger* herself could not have carried all that was collected for the *Quest*, with room to spare for its efficient handling. But the plan was great and its accomplishment still practicable if circumstances should prove as favourable as they might just possibly be. The chance of success was sufficient for Shackleton, who went about his preparations with the heart of a boy, though old friends saw in his face signs of the wear and tear of his long years of unceasing hardship and toil.

The *Quest* after being refitted at Southampton came into the Thames on 15th August to complete her equipment. Shackleton had been elected to the Royal Yacht Squadron, and his little vessel flew the white ensign and the burgee of the premier yacht

THE QUEST, UNLOADED.

Face p. 272.

club of the world. The King had received the explorer and presented a silk Union Jack to be carried on the expedition. Queen Alexandra received him and heard the new plans with all the friendly solicitude she had shown for every Antarctic expedition since she saw the *Discovery* at Cowes in the first year of the century. The work on board was much hampered by visitors, and the stowing of the hold was a troublesome task; but light hearts and willing hands got the ship ready in St. Katharine's Dock.

On 18th September the little vessel left the dock, passed under the Tower Bridge and made her way down the Thames, as the *Nimrod* and the *Endurance* had done before her. There was much popular interest along the river-side, and the shipping gave a noisy send-off to the expedition. On board there was scarcely standing room, with a crew of equals not yet shaken into shape, a crowd of friends, unstowed stores and open hatches. In the midst of the excitement Shackleton discovered that the black kitten presented to the expedition for good luck was starving. " Bring the milk ! " he ordered; but there was no milk in the galley, where black coffee was being served. " Get up a case of milk from the hold ! " a member of the expedition disappeared and in time heaved up a box on to the crowded deck. " Break it open ! " roared Shackleton, and the box was smashed up. " Open a tin and feed the cat ! " Possibly words of emphasis have been omitted, but the whole scene reminds one in spirit and setting of Nanty Ewart on the *Jumping Jenny* in *Redgauntlet*, ordering tea for his sea-sick passenger. So little change do the centuries bring in the heart and the language of the true sons of the sea.

The *Quest* took her departure from Plymouth on 24th September, all on board in the highest spirits. Shackleton felt in the sea-air the breath of freedom after his years ashore and in the unnatural luxury of passenger vessels, and he shook off the hateful ways of business, which had always placed him at a disadvantage, for the delights of command, where he was supreme alike in his power and in the unquestioning devotion of his crew. It was as remarkable a crew as ever the white ensign waved over. It represented almost every portion of the Empire, eighteen

British subjects all told, but only one in the narrow geographical sense an Englishman, the rest were Scots, Irish, Canadian, Australian, New Zealanders. Five of them had been with Shackleton on the *Endurance*: these were Frank Wild, as before, second in command; Frank Worsley, once more sailing master; A. H. Macklin and J. A. McIlroy, the surgeons; L. D. A. Hussey, meteorologist; and C. J. Green, the cook. Two were boy scouts chosen by the leader out of ten selected by Sir Robert Baden-Powell as the best of a thousand applicants. One, G. Wilkins, naturalist, had previous experience in the Arctic regions on Canadian expeditions. Almost all had naval or military rank as officers during the war; now they were on the same level doing all the work of the ship, each man putting his hand to any job irrespective of his title or designation. As a temporary member of the party, Mr. G. S. Lysaght, Shackleton's old friend, a hardened yachtsman, came as far as St. Vincent.

Only the best of good fortune could possibly make the expedition a full success; every day was required to reach Cape Town in time for the attack on the Enderby Quadrant. Only the best weather could enable so small a vessel, so hastily equipped, to cross the great stretches of the ocean without loss or damage to the complicated gear crowded on board her. Shackleton had never left home with more odds against success in his adventure; he knew the risk, but he saw the one chance and ventured it.

For two days the *Quest* had fine weather, but showed herself a lively ship. She was very low in the water, and had a great double bridge and deckhouse, tall masts, and long yards on the foremast, so that once set rolling she kept it up, and when the wind freshened and the sea rose she took the water over her bows in fine style, and kept it washing about the deck. Still she was a good sea-boat and, handled skilfully as she was, she was safe though terribly uncomfortable For a week she met strong head winds, her engine developed defects, and the rigging was strained, while only Shackleton, Lysaght, and two others escaped sea-sickness. All the rest were ill, two so seriously that they had to return from Madeira The exceptional severity of the weather and the unexpected defects in the ship taxed

Shackleton's strength severely, as for five days he never left the bridge except to go below to cheer up the invalids. The necessary repairs were carried out in a week at Lisbon, the *Quest* having struggled into the Tagus on 4th October. The voyage to Madeira was carried out in still worse weather, perpetual head winds, but there was a pleasant run through the Trades to St. Vincent, which was reached on 26th October. Here Mr. Lysaght left with great reluctance, after a farewell feast the *menu* card of which contained verses by Shackleton as a tribute to his friend, including these lines:

" You have watched our reeling spars sweep past the steady stars
      In the storm-wracked night.
You saw great liners turn ; high bows that seemed to churn
      The swell we wallowed in.
They veered from their ordered ways, from the need of their time-kept days
      To speed us on "

During the two days at St. Vincent, Shackleton was struck down with what he called suppressed influenza ; but he picked up on the voyage westward to the lonely islet of St. Paul's Rocks, which he had sighted from the *Hoghton Tower* thirty-one years before, and where he now began to carry out the programme of his last expedition by sending the naturalists ashore to make collections. He had been obliged by the bad weather to change his plans, and to omit the call at Cape Town (though his seaplane and the polar equipment had been sent on there from Madeira to await the *Quest*). Rio de Janeiro was to be the last port in touch with regular mail-boats.

He reached Rio on 22nd November and was most courteously treated by the Brazilian Government, which gave him the free use of the naval dockyard, and allowed a spar to be taken from one of their men-of-war to replace the strained foretopmast of the *Quest*. The faulty engines were got into good running order, and the anxiety as to the fitness of the ship for her task was relieved. But the heat was extreme, and Shackleton suffered from it. It was observed that his spirits flagged, and that he sat almost silent through some of the banquets given in his honour. But on St Andrew's night he gave a speech in his own inimitable style at the Scottish gathering, and on 5th December he roused himself to deliver one of those wonderful lectures which had given such pleasure to audiences in all parts of the world, and cast so warm a glow of romance on the bare facts of exploration. The Municipal Theatre was filled to hear him, and the whole diplomatic corps, even the representatives of the ex-enemy States, supported the high dignitaries of Brazil in the audience, and, as of old, he charmed them ; but the effort shook him.

To Dame Janet Stancomb-Wills, who had never ceased to sympathize with his aspirations and to promote his plans since the old days of the *Endurance*, he wrote :

"We are tucked away about nine miles from Rio on an island, baking hot, with mosquitoes and clanging hammers all day ; then at night I have to dress and attend some long-drawn-out function, and all the time I am mad to get away. . . . From South Georgia we go into the ice, into the life that is mine, and I do pray that we will make good. It will be my last time. I want to write your good name high on the map, and however erratic I may seem, always remember this, that I go to work secure in the trust of a few who know me"

Sighing for the health and happiness which he believed awaited him in the far South, he sailed from Rio on 19th December, the forts saluting his departure. Just before leaving he wrote to Mr. Rowett :

"110° in the shade ! All the work is done, and we are going The next you will hear will be, please God, success. Should anything happen in the ice it will have nothing to do with anything wrong with the ship *The ship is all right.*

> " Never for me the lowered banner,
> Never the lost endeavour !

" Your friend, ERNEST."

Outside there was no rest nor peace. A great albatross met the *Quest* as a convoy a few degrees south of the tropic, farther from the South than any of its kind had been known to stray before. The voyage southward was hampered by contrary winds, rising on Christmas Eve to the most frightful storm in all Shackleton's experience, in which the *Quest* had to lie hove-to for a whole day, using oil to check the breaking of the sea, and rolling 50° to each side as she rose and sank 40 feet on the crest and into the trough of the tremendous waves. No cooking was possible and no rest. But the stout little craft weathered the storm, and the wearied leader, casting his mind back over all his past experience, acknowledged that never before at sea had he so longed to reach port.

Calm came with the New Year, and on 2nd January the first iceberg swam into their ken, glorious with submerged spurs of glittering green, and caverns of deep azure. It brought a message of welcome and of hope; yet he wrote of it:

" The old familiar sight aroused in me memories that the strenuous years had deadened, Ah me! the years that have gone since, in the pride of young manhood, I first went forth to the fight. I grow old and tired, but must always lead on."

On 4th January 1922 the *Quest* reached South Georgia and cast anchor off the Grytviken whaling station. Shackleton went on shore at once to see old friends and make preparations for getting the ship ready with the least delay to proceed on the much-reduced Antarctic cruise that the late date permitted. In the evening he returned on board apparently quite well. He told his comrades that he was now happy and contented, feeling that the voyage had really begun. He sat down cheerfully to write his diary, and the last words were: " In the darkening twilight I saw a lone star hover gem-like above the bay."

And with his last thoughts of " sunset and evening star " the Leader went to rest; but he could not sleep, and in a few hours came the " clear call " which made the crossing of the Bar a matter of moments. He died in an attack of angina pectoris at 3.30 a.m. on 5th January.

A fine, a characteristic end, without warning, without regret. Life stopped in the course of a new onward movement. All his life had been a rattling rush of swift succeeding action, like a chain cable racing through the hawse-pipe into an unfathomed sea, causing the world to vibrate as it ran out its full length of forty-seven shackles when the last link slipped over, and there was silence.

Only Dr. Macklin, who happened to be on watch, was present at the closing scene, which passed so quickly that death came before medical help could be given. Wild, McIlroy, and Hussey were called, but all the rest of the ship's company slept until daylight, when the heavy news was told There was a day of gloom and consternation unknown before in any ship or camp of his.

Wild succeeded to the command, and decided that the expedition should go forward, knowing that this would have been the wish of the leader. The wireless installation had broken down, and no message could be sent out till the next ship sailed. The body was embalmed and removed to the little Lutheran church, as it was decided to send it home for burial unless other instructions were received at the first port. His old comrades thought that burial in South Georgia would have met Shackleton's own wishes best; but they wisely felt that Lady Shackleton should be consulted before the step was taken. It was decided that one trusted friend should accompany the body and see the desired arrangements carried out; the choice fell on Hussey.

On 19th January the body was placed on the Norwegian steamer *Professor Gruvel* and the long funeral procession began. The albatrosses followed in the wake of the ship to the limit of their range, watchful and inscrutable as ever, bringing the memories and the mysticism of centuries with them. In ten days the ship reached Montevideo, and next day the death of Shackleton passed like a shadow over the world. The sorrow of nearest friends could not but be softened by the altar smoke of sympathetic appreciation of the bright and vivid life, which rolled in from the highest and the humblest in every land through which he had flashed; and where had he not been?

At Montevideo the Uruguayan Government made the occasion one of national mourning, and accorded to Shackleton the full honours due to a Minister of State. A hundred marines were marched down to the ship's side to bear the coffin, covered by the British flag, to the military hospital, where it lay in state under guard. By one of the happy coincidences that we have had to notice so often in these pages, the commander of the guard of honour was the captain of the *Instituto de Pesca*, in which the second attempt for the relief of the Elephant Island party had been made.

With unerring instinct Lady Shackleton decided that the burial should take place in South Georgia, under the shadow of the mountains he had been the first to cross. On 15th February, after a ceremony in the English Church attended by the President and members of the Government, the diplomatic body and deputations from the British communities of Uruguay and the Argentine, Shackleton's remains were carried with all the pomp that Latin America could muster through streets lined with national troops to the British ship *Woodville*, bound for South Georgia. The Government had offered to send the remains south on one of their warships, but this offer was declined on account of the risk of such a vessel being unable to penetrate the ice which was apt to be encountered round the sub-Antarctic island; but the cruiser *Uruguay* accompanied the *Woodville* to the limit of territorial waters, and fired a salute of nineteen guns in honour of the dead.

So Hussey brought back his leader to the region he loved so well, and with the escort of the ever-watchful albatrosses, like a cloud of witnesses of the great sea-leaders of the past, the funeral procession of more than 3000 miles terminated at Grytviken. The community who lived at the whaling stations along the coast was small, and not of our race, but the Norwegians had taken Shackleton to their hearts long before, and treated him as a true descendant of the Sea Kings. When the last service was held on 5th March, half in Norwegian, half in English, in the little Lutheran church at Grytviken, the managers even from the most distant stations came to attend it, and one of them, with a fine and understanding sympathy, brought with him from a great distance seven ex-service men from a British ship, Shetlanders all, to act as bearers on the last stage. Though Hussey was the only one of his own " boys " there, he felt that the little crowd of rough sailors,

280 THE LIFE OF SIR ERNEST SHACKLETON

and the one lady who lived on the island and laid flowers on the grave, were all true mourners

On 2nd March the greatest of the many memorial services held at home took place in St. Paul's Cathedral, where a vast congregation assembled, including the family, representatives of the King and Queen, of Queen Alexandra and the Prince of Wales, of the Admiralty, the Royal Geographical Society and other public bodies, old comrades who had sailed with Shackleton on the *Discovery*, the *Nimrod*, the *Endurance*, and the *Quest*, business friends, social friends, and the great general public. The service, faultless in its beauty and magnificent in its setting, concluded with the sounding of the Last Post by cadets from the Thames Nautical College; but the minds of many present were away in the grander scene at South Georgia under the dome of heaven, with the wash of the waves and the wailing of the sea-birds as the most plaintive of all funeral music.

Seldom indeed can all that is most moving in pathos and pageantry have been gathered into a funeral so prolonged, and the man who was buried at Grytviken was the boy who at Kilkea, in Dublin, and in Sydenham, never tired of following funerals, as if that were the chief end of life

How the character of the man impressed his followers, "the boys," who held him as their hero, has been partly shown in the several chapters, but appears even more clearly in the personal experience of one who sailed in the *Endurance* and the *Quest*, Mr. L. D. A. Hussey:

"What appealed to me more than anything was the fact that he was so human. He had his faults, and knew it too, and he expected perfection in no man; but he was quite willing to overlook what was bad, and just remember the good in every one. He had a way of compelling loyalty. We would have gone anywhere without question just on his order. And his personality left its mark on all our lives. Sometimes we would be inclined to do a thing which we knew he would not approve of, and, though there was no possibility of its ever being found out, the mere fact that we knew that he would not have liked it, proved a very efficient check. Even now our thoughts and ideas are all coloured by what we knew of his. We are constantly, and almost unconsciously, saying to one another, "The Boss would never have done that," or "That is just what the Boss would do." In this way our lives still seem to have some reference to him, and his ideas and aims are set up as our ideals. Now that he has gone there is a gap in our lives that can never be filled. He was the best friend that we have ever had, and in attempting to approach his ideals, and in always associating him with those ideals, we believe that he is attaining the immortality that he desired."

Another comrade of the *Endurance*, Mr. J. M. Wordie, contributed a feeling obituary notice to the *Geographical Journal*, and we quote a portion of it, revised and amplified by him:

"Shackleton possessed in unusual measure the highly poetic imagination which is traditionally associated with a love of exploration. It is well expressed in his writings and in the naming of his ships; still more in his love of poetry. His wonderful memory made it easy for him to have ready a line of verse suitable to almost every occasion. It would generally be from Browning, his favourite poet. When combined with great physical strength and with powers of leadership, a poetic nature such as Shackleton's is the very stuff from which the greatest explorers are made.

"Shackleton, indeed, possessed the faculty of leadership to a pre-éminent degree. That, together with his generosity, made all the best men who had served with him his staunch adherents. They had implicit faith in his judgment. Shackleton's was the rarer type of courage which is controlled rather than rash. Two of his greatest decisions—to turn back on the southern journey in 1909, and to remain on the ice after the *Endurance* was crushed in 1915—are examples of this. Caution and shrewdness were combined, however, with invincible optimism; this made him a trying partner at card games, and was also responsible for a continual hankering after and belief in hidden treasure. The latter feature was but another instance of his romantic nature. It was perhaps this which first suggested to his intimates a likeness to Raleigh. Then his friends found that he was a Raleigh in many ways—courtier, poet, explorer, and lover of his country. I always pictured Shackleton as very much the Elizabethan. We once had an amusing time together discussing the voyagers of that period, and trying to reconstruct the life and behaviour on board the ships, the aims, ambitions and characters, even the garb of the officers and crews. In an age which is producing modern Elizabethans, Shackleton will surely be reckoned as most true to type "

When the *Quest* came back to Grytviken after her cruise in the Antarctic, Wild and his comrades heaped together a great cairn, surmounted by a cross and cemented into solid rock, as a memorial of their leader Since the road of loving hearts was hewed by Stevenson's native friends to his mountain grave in Samoa, no memorial has been put together more notable in its tribute of devotion than this rough pile untouched by a hireling hand.

MEMORIAL CAIRN AT GRYTVIKEN.

Face p. 282.

EPILOGUE

' Thronging through the cloud-rift whose are they, the faces
Faint revealed but sure divined, the famous ones of old ?
' What '—they smile—' our names, our deeds so soon erases
Time upon his tablet where Life's glory lies enrolled ?

' Was it for mere fool's play, make-believe and mumming,
So we battled it like men, not boy-like sulked or whined ?
Each of us heard clang God's " Come ! " and each was coming :
Soldiers all to forward face, not sneaks to lag behind ! ' "
ROBERT BROWNING.

EPILOGUE

WE believe that the foregoing chapters show that Shackleton possessed many of those qualities which detach a man from the background of his contemporaries to be silhouetted for ever against the sunset sky of time. And by the foreshortening power of history he will appear there side by side with the great adventurers of all the centuries. He drew his inspiration mainly from the Victorian poets, but in his maturity he would have felt himself an alien in the conventionally ordered life of the nineteenth century. In the twentieth century he was more fitly placed for winning appreciation, but to set him amongst his peers in his true environment we must spin the globe backwards through at least three hundred years.

It is easy to picture Shackleton strutting in all the bravery of Elizabethan finery along the miry quaysides of Wapping, and outdoing Raleigh in the graceful casting of his costly mantle before the feet of his queen. No effort is required to see him " beating about the streights " in a consort of the *Golden Hind*, or matching Drake at bowls on Plymouth Hoe, though we think we see him rushing from the Hoe to the harbour at the alarm of the Armada—his ship would have been the first to be ready for the fight. How his spirit would have leaped with Frobisher's at the lure of unmined treasure in the farthest north! We can see him risking everything to load his ship yet deeper with the glittering treasure that promised wealth but proved so worthless, and we can see him emerging from his bitter disappointment to seek still more perilous adventures. The possibility of an ancestor in common between Frobisher and Shackleton allows us to linger in the pleasant dream that both polar explorers drew their lust for adventure from the same

286   THE LIFE OF SIR ERNEST SHACKLETON

origin. Whether the insubstantial fabric can sustain the bond of blood or not, the spiritual affinity stands.

And how the heart would leap within him during his evenings at the Mermaid !

If Shackleton had lived in the sixteenth century he might well have been knighted on the quarter-deck of his ship, and in the greater swing of chances in those spacious days he might equally well have been beheaded in the Tower. He would certainly have cut a dashing figure and made devoted friends and bitter enemies amongst the great men of that day.

In modern England, Shackleton had less chance of being understood, for he did not fit into any of the ready-made clothes by which we are in the habit of classifying people, and we could search in vain amongst the printed labels for one to describe him. His life was like a mighty rushing wind, and the very strength of his nature made him enemies as well as friends. In many ways he was always a boy, and that boy was often possessed by a spirit of pure mischief. When he found any one shocked at his disregard of conventional decorum, he was sometimes drawn on to see how much his victim could stand, and this led to unfortunate misunderstandings, though no harm was meant, and no one with a sense of humour was ever hurt. After such a piece of fun Shackleton was usually sorry.

He sometimes evoked dislike without making an enemy. For example, one of those who did not like him complained that Shackleton always seemed to be on the defensive in his company. As the foregoing chapters show again and again, Shackleton, with all his strength, was extremely sensitive, and he adjusted himself instinctively to his environment: feeling an atmosphere of challenge, he would assume the defensive, not because it was his natural attitude, but as the inevitable response to suspicion. And men of science have misinterpreted such instincts, even in the lower creatures, like the French naturalist who wrote " cet animal est bien méchant, quand on l'attaque il se defend "

To sympathy Shackleton responded like a flower to the sun, and all through his life his deepest self was revealed in the company of warm-hearted and congenial friends. In South America, when smiles were the only common language, his hostesses would find a method of telling him in Spanish that he was *simpatico*.

His character was compounded of such contrary elements that he had points of contact, and even of sympathy, with people of every kind. With one he would have a boxing bout, caring only that his fellow was a good boxer, no matter how big a scoundrel; with another, a discussion of the beauties of the book of Job, caring only for his poetical perceptions, even though he were an archbishop. He could carouse with sailors, drinking out of tin mugs at a dirty table and find nothing wanting; while at a fine dinner, amongst the perfection of silver, glass, and china, a chance stain on a knife-blade might rouse him to disgust. He held a horse for a quarter of an hour at the call of a farmer in a little town in Canada, because the man was transparently honest and unconscious of asking anything out of the way. And, when in one of the most exclusive hotels of one of the greatest capitals of Europe the hall porter failed in civility, Shackleton seized him by his gold-laced collar and ran him across a crowded hall to the office of the manager, and extorted apologies from both because he would not stand calculated insolence. Every one who knew Shackleton can cite many cases of such apparently contrary impulses and actions, and it is no wonder that people who cannot admit the appearance of inconsistency in their friends should dislike him.

The way to understand his character as we conceive it is to take account of the scale on which it was fashioned. The man was great and had to be taken as a whole. Most people from ignorance of his size and diversity, and from want of experience of men of his magnitude, only saw a small part of him, liked it, and hailed him as friend; disliked it, and perhaps held their peace. Viewed in its full size, the strength of the whole fabric made criticism of the finish and polish of its parts superfluous and impertinent. As a great steel girder may shine in the sunlight and yet be flecked with shadows which may hide defects that would ruin a slender tube, so Shackleton's character had its shades which in a small man might be serious, but in him could be viewed as mere accidents of his strength

Shackleton was human before everything; he had virtues and faults which we have tried neither to exaggerate nor to conceal. He was often pleased with himself and often disappointed. He was essentially a fighter, afraid of nothing and of nobody, and every one with whom he crossed swords did not like him the better for it. People have said that he was vain and very open to flattery, but much as he liked to feel himself appreciated, he had a keen scent for insincerity, and of all his hatreds that of a hypocrite was the greatest. He was intolerant of restrictions, and officialism of every kind was repugnant to him. In a sense he was a law unto himself, and made no concealment of the fact that he was never really happy unless he was absolutely free.

He overflowed with kindness, and could never bear to see any one unhappy or in want without offering help. So deeply did the pangs of hunger burn themselves into his memory on the *Discovery* expedition, that he made a habit of feeding every hungry person he met, especially children. It was no unusual thing to see a group of hungry street waifs at the door of a tea-shop near his office in Regent Street waiting for him to pass by, and he never failed to take them in and feed them. A coffee-stall near Hyde Park Corner which he often passed in the dark evenings when going home from work, was another scene where many a hungry wanderer was warmed and fed by him. He had a cheery word for all he met. If he had to wait in an office for the principal to be free, he went round and chatted to the clerks and porters, whose good will could never be of any use to him. It was his instinct to spread kindliness. His disposition was trustful and affectionate to all his friends. As a son, as a brother, as a husband and a father he showed a wealth of love and tenderness that could not be surpassed

At the commencement of every voyage all through his career he suffered acutely from home-sickness, yet the wander-fever was not checked and the master-passion of his life was probably the outward urge into the unknown. It was a primal impulse.

EPILOGUE

He often tried to explain, but with little success, the inducements that drew him away from the comforts of home and society. It was his nature to push into the unknown, to take great risks, to make quick decisions and sudden changes of plan. His nature impelled him always to be doing things, especially difficult things, preferably dangerous things, above all big things, and to be doing them with all his might; the greater the obstacles the happier the endeavour No one ever exemplified better the pure romance of exploration The question he should have been asked was, Can you ever rest? for rest was to him the abnormal state which, when it supervened, had to be accounted for. The constant activity wore him out. In his all-but forty-eight years he had done the deeds and experienced the emotions of a long life. All his rest was left to a possible future when, we may perhaps deduce from the lives of his ancestors, a dormant homing instinct might emerge to draw him back to the land with the placid interests of field and garden.

Shackleton sometimes did things which were wrong when tested by the standards of ordinary people; he made rash promises which he ought not to have made, which circumstances sometimes compelled him to break. His generosity often outran his means, and there was nothing he possessed that he would not give away. He did not verify quotations, was careless as to dates, and liked to tell stories that were true in a large poetic sense. But he knew his failings, and his fights within himself were perhaps as hard as those with external difficulties. Viewed in the proper perspective, he was successful; that is to say, in the struggles of his constant striving he made more successes than failures.

The nearest we can come to an understanding of Shackleton's aim in life is that he was determined to deserve success and to attain it. He enjoyed honours, but only because he felt they were deserved. He strove to do great things, not because he had an overpowering passion for the apparent goal, but because he burned to do his best—the prize was in the process. His great ambition was to follow his ideals worthily. He was not the champion of causes so much as of methods.

Like an abstract mathematician—a creature, his very antipodes in nature—his interest lay in working out a problem in the best possible way, though what x, y, z meant was nothing to him, unless possibly he permitted a glimmering hope that they might one day resolve themselves into £ s. d.

Of course Shackleton made leeway while beating to windward across the ocean of life; but he kept his eye on the compass, and he never let go the helm. He looked on courage as one of the minor virtues, for he took it as a matter of course that every true man is courageous. Endurance and perseverance he ranked higher, and practised almost to the attainment of perfection To others the strength and sweetness of his personality appeared to be his dominant quality. His quick, Irish wit played over the surface of his Yorkshire caution, and together they went far to explain the magnetic charm of manner which he never lost. Optimism was perhaps the quality he set most store by, but it was not the slipshod optimism of Mr. Micawber passively waiting for something to turn up, not the cynical optimism of Candide, with no higher teaching than the blessedness of cultivating a garden His optimism was that preached by Robert Browning, and the event almost suggests that he set himself to live the life of faith in the essential goodness of things, as Browning taught it To strive and thrive, to fight on and hope ever, to play a great part in a world with a happy ending, were living motives for him.

Quite characteristically this whole-hearted absorption of Browning's philosophy, and his immense knowledge of Browning's works, did not lead Shackleton to neglect or despise other poetry. His taste was singularly wide He revelled in the voluptuous alliterations of Swinburne in his earlier days. He liked the rough realism of Masefield's sea poems and the clang of Kipling's ballads; most of all, he loved the call of the wild, as given out by Robert Service; but he kept his early liking for Tennyson, and gloried in Milton and Shakespeare. But Shackleton was human above all things, and could enjoy a music-hall ditty, or a deep-sea chanty among "the boys" with unaffected pleasure, and without the least feeling of incompatibility, he would end an evening of the wildest jollity with "Lead Kindly Light," or "Fight the Good Fight." The depth of his consistency could never be plumbed by the world's coarse thumb and finger.

In Shackleton the religious sense was strong, though he could hardly be said to conform to any of the recognized modes of expression. He believed the soul to be immortal, and was very sure of providential guidance. His God was the God of Nature, of the stars, the seas, and the open spaces, of the great movements of history and the abysmal depths of personality. But no creed going beyond the bottomless words, "I believe," could contain a definition of his faith There was goodness permeating Nature, and the world was progressing towards good. So he believed, and hence it is difficult for those who hold by forms and articles to realize that he had the vision of Truth; and impossible for the school of thought, which sees only disgrace in the past and gloom in the future of human endeavour, to understand the ground that Shackleton stood on.

For many years the poem "Prospice" had a special meaning for Shackleton and his wife They used the word in telegrams as a code symbolical of hope; and no fitter close for this little book on a great life can be found than the lines :

"I was ever a fighter, so one fight more,
    The best and the last !
I would hate that death bandaged my eyes, and forebore,
    And bade me creep past

No ! let me taste the whole of it, fare like my peers
    The heroes of old.
Bear the brunt, in a minute pay glad life's arrears
    Of pain, darkness, and cold;

For sudden the worst turns the best to the brave."

APPENDIX

Compiled by Lady Shackleton

LIST OF DISTINCTIONS CONFERRED ON SIR ERNEST SHACKLETON

I. BY HEADS OF STATES

DECORATIONS, BRITISH

Member of the Royal Victorian Order, 1907.
Honour of Knighthood, 1909.
Commander of the Royal Victorian Order, 1909.
Officer of the Order of the British Empire (Military), 1919.
British War Medal for the Great War, 1919.
Victory Medal with emblem for Mentions in Dispatches, 1919.
Arctic Medal with three clasps, 1904, 1909, 1917.

DECORATIONS, FOREIGN

Polar Star of Sweden, 1909.
Dannebrog of Denmark, 1009.
St. Olaf of Norway, 1909.
Legion of Honour (France), 1909.
St. Anne of Russia, 1910.
Crown of Italy, 1910.
Royal Crown of Prussia, 1911.
Order of Merit (Chile), 1916.

II. BY SOCIETIES AND CORPORATIONS

SOCIETIES, ETC., BRITISH

Trinity House, Election as Younger Brother of, 1909.
Royal Geographical Society. Silver Medal, 1904.
Royal Geographical Society · Special Gold Medal, 1909.
Royal Scottish Geographical Society Silver Medal, 1906.
Royal Scottish Geographical Society. Livingstone Gold Medal, 1909.
Tyneside Geographical Society: Gold Medal, 1909.
University of Glasgow · Honorary Degree of LL.D., 1914.

SOCIETIES, ETC., FOREIGN

Austria . . Geographical Society of Vienna: Honorary Membership, 1910
Belgium . Royal Belgian Geographical Society, Brussels: Gold Medal and Honorary Membership, 1909.
Royal Geographical Society of Antwerp Gold Medal, 1909.
Royal Yacht Club, Gold Medal, 1909.
Brazil . . Aero Club; Honorary Membership, 1921.
Chile Life-saving Society of Valparaiso : Gold Medal, 1916.
Patriotic Military League, Santiago: Special Gold Medal for Discipline and Valour, 1916.
Scientific Society of Chile, Santiago: Honorary Membership, 1916.
Municipality of Punta Arenas Gold Medal, 1916.
Denmark . Royal Danish Geographical Society, Copenhagen · Gold Medal, 1909
France . . . Society of Geography, Paris: Gold Medal, 1909.
City of Paris . Gold Medal, 1909.
Academy of Sciences, Paris Prix Delalande-Guérineau for Geography, 1916.
Germany Geographical Society of Berlin . Gold Medal, 1910.
Geographical Society of Frankfurt · Ruppell Gold Medal and Honorary Membership, 1910.
Italy . . Royal Italian Geographical Society, Rome · Gold Medal, 1909
Norway . . Geographical Society, Christiania: Honorary Membership, 1909.
Russia . . Imperial Geographical Society, St. Petersburg: Gold Medal, 1910.
Geographical Society of North Russia, Archangel: Honorary Membership, 1919.
Sweden . . Geographical Society, Stockholm · Gold Medal and Honorary Membership, 1909.
Royal Science and Literary Guild, Gothenburg: Honorary Membership, 1909.
Switzerland . Society of Geography, Neuchâtel Honorary Membership, 1910.
United States . American Geographical Society of New York: Cullum Gold Medal, 1909
American Museum of Natural History, New York: Honorary Fellowship, 1917
National Geographic Society, Washington · Hubbard Gold Medal, 1910.
Geographical Society of Philadelphia: Elisha K. Kane Gold Medal and Hon. Membership, 1910.
Art Club of Philadelphia: Hon. Membership, 1912.
Geographical Society of Chicago Helen Culver Gold Medal, 1910.

INDEX

Aberdeen lectures in, 85, 172, 266, sails from, 256; House, Sydenham, 22, 49
Abraham, John (ancestor), 7.
Actor, powers as, 178–179
Adams, Lieut Jameson Boyd, R.N.R., 97, as meteorologist, 108, 125, climbs Mount Erebus, 129; on Southern journey, 134, 135, 149, 151–153, gets medal, 162
Adare, Cape, discovery of, 15; first landing, 48, 52, landing, 63; 152
Address to electors at Dundee, 94; recruiting, in Sydney, 248
Adelaide, lecture at, 158.
Adélie Land, 14, 190.
Admiralty . decisions of, 82, 202, 230; distrust of, 250; help from, 110, 234.
Adventure : love of, 27, 99, 199, 260, 285–9; at Tocapilla, 41.
Advertising himself, charges of, 187, 271.
Affection, 21, 288
Agent experience as, 91; use of, 109, 122, 167, 175.
Albatross chickens, 226
— mystic associations of, 36, 42, 152, 205, 225, 276, 278, 279.
Albemarle, H.M.S , 107
Albert Hall, lectures at, 161, 265
Aldrich, Admiral Pelham, 83
Alexander I , Tsar, 12
Alexandra, Queen, on *Discovery*, 60; on *Nimrod*, 110; on *Endurance*, 201, received by, 162, 166, 251, 273
Allardyce Range, crossing of, 227
Alligators as pets, 47
Altitude, effect of great, 143
Ambition, personal, 50, 51, 52, 212, 289
America, visits to, 47, 48, 49, 80, 172, 189, 190, 249.

American business projects, 177, Geographical Society, 173; humour, 176; Museum of Natural History, 249
Amundsen, Capt. Roald, 108, 153, 166, 184, 186, 188
Ancestors, 3–10; spirits of, 225
"Ancient Mariner " 98, 225, 226.
Andorra, sailor from, 233
Ann Arbor, lecture at, 175
Ann, Cape, 13.
Antarctic : in Ross Sea, 46 , in Weddell Sea, 60; loss of, 83, 211.
Antarctic Circle, 3, 7, 12, 13, 15, 39, 42, 49, 115, 206, 215.
Antarctic Continent · search for, 7, 16 , first on, 46, first wintering on, 52
Antarctic Days, Preface to, 191
Antarctic exhibition, 165.
Antarctic expedition : early, 11–16, 39, 42 , Belgian, 49, 52; National, 54, 56–80; schemes for, 98; British (1907), 103–155, proposed Austrian, 194, Imperial trans-, 194; Shackleton-Rowett, 271
— Regions, maps of, 11, 196
— stamps, 116, 144.
Antarctique, Mère, 167.
Appreciations by friends, ix, 51, 56, 84, 98, 130, 180, 217, 247, 261, 278, 280, 281
Apprenticeship . sea, 34; polar, 74
Archangel, visit to, 257, 263
Archbishop, value of, 255, 287.
Arctic Circle crossed, 256
— Exploration, 21, 46, 177, 268, 269.
Arctowski, Dr H , 105
Argentine, visits to, 204, 240, 254
Armadale Castle, 163.
Armitage, Cape, 68, 127, 149
— Capt Albert B., R.N.R , 58, 64, 71, 158
Arms of Shackleton, 8

295

296　　　　　　　INDEX

Army Balloon Department, 60
Armytage, Bertram, 126, 132, 162
Arnold, Matthew, quoted, 1, 193.
Asquith, Right Hon H H., 164
Athlone, Earl of, 265
Aurora, polar ship, 190, 195, 205, 229, 230, 238, 242, 244
Aurora Australis, polar book, 129
Australia, visits to, 39, 112, 158, 245, 248
Australian Antarctic Expedition, 185 ; Government, 112, 238, 248 ; Relief Committee, 238, 245
Austria, visits to, 170, 179.

Baden-Powell, Sir Robert, 274
Bahia, visit to, 253
Bailey, P J, quoted, 114.
Bâle, lecture at, 179
Balleny Islands, 13, 230
Balleny, John, 13
Ballitore, Co. Kildare, 5, 6, 17, 36.
Ballitore Papers, 6.
Balloon on *Discovery*, 60, 66, 105
— Bight or Inlet, 66, 117
Balmoral, lecture at, 165.
Baltimore, lecture at, 249.
Banjo, 237, 266
Banshee, Irish packet, 22, 37
Barbados, visit to, 253
Barne, Lieut Michael, R N , 58, 72, 105, 107, 133
Barrier, Ross or Great Ice · discovery of, 15 ; Borchgrevink on, 54 , sea-face of, 64, 65, 116-120 , travelling on, 76, 128, 133-138, 146-149
Bartholomew, Dr John George, 86
Bartlett, H H , 105
Bass Strait, 39
Beagle, H M.S., 238.
Beardmore, William (Lord Invernairn), 88, 92, 97, 103
— Glacier, 139-142, 145, 146, 151, 189, 230, 243
Bears in Antarctic, Kaiser on, 171
Beaufort Sea exploration, 268, 269, 270.
Beaufoy, cutter, 12
Beaumont, Admiral Sir Lewis, 168, 230
Becker, Sir Frederick, 270.
Bees, instinct of, 54
Belfast, lecture at, 168
Belgians. Leopold II , King of, 93, 167 , Albert, King of, 167.

Belgica in Antarctic, 49, 50, 52
Bell, Captain William and Ellen, 8.
— Dr William, 19.
— William, jun , 105.
Bellingshausen, Admiral, 12.
Bengal, Bay of, 39
Berlin, 201 , lectures at, 169, 170.
Bernacchi, Louis C , on *Discovery*, 58, 63, 70
Berwick, H M S , 97.
Bird, Cape, 127.
Birth, 3, 10 , of eldest son, 91 ; of daughter, 103 ; of second son, 185
Birthday: on *Discovery*, 67 ; on Barrier, 148.
Biscoe, John, 13
Biscuit, gift of, 146
Bjorn, inspected, 108.
Blackie, Prof John Stuart, 88.
Blizzard, The, magazine, 71.
Blizzards, 121, 129, 133, 143, 144, 148, 149, 189, 243.
Blubber stove, 219.
Bluff, The, N.Z , 155
Bluff depot : (1909) laying, 135 ; finding, 149 ; (1916), 243
Board of Trade marine department, ix ; exam., 43, 48.
Boats, working, and at sea, 33, 150, 220, 224, 226, 255.
Bohemian Club dinner, 249
Bonaparte, Prince Roland, 168
Books read at sea, 35, 46, 48, 61, 62, 71
Borchgrevink, Carstens Egeberg, 46, 48, 52, 54, 65, 117, 153.
Boss, The, nickname, 130, 266.
Boston, lectures at, 173, 249
Bowers, Lieut Henry R , 189
Boxing, love of, 32, 259, 287.
Boy, rating of, 30.
Boyishness, 99, 160, 269, 271, 286
Boy Scouts, 256, 274
Brandon, lecture at, 177.
Bransfield Strait, 218
Brazil. sighted, 37 ; visited, 253, 275.
Breslau, lecture at, 171.
British Antarctic Expedition, 1907, 105-155 ; small expense, 164
— Association meetings, 83, 89.
Broads, Norfolk, visit to, 192.
Brocklehurst, Sir Philip, Bart. : on *Nimrod*, 108, 126 ; climbs Mount Erebus, 129 ; illness of, 130 ; gets medal, 162.
Brocklesby, Dr. Richard, 6.

INDEX

Brown, Prof. A. Crum, 88.
— C. P., of Valparaiso, 42
Browning, Robert : poetry of, 49, 50, 51, 56, 63, 66, 70, 281 ; quoted, iii, 101, 103, 156, 181, 283, 291
— Settlement, 163, 199, 266
Bruce, Dr William Speirs, 60, 89, 184, 193, 207, 209, 231.
Brussels, reception at, 167.
Bryant, Henry G , 173.
Bryce, James (Viscount Bryce), 173.
Buckley, George, 105, 114, 116
Budapest, visits to, 170, 186.
Buenos Aires, 203, 240, 253.
Burke, Edmund, 6, 7, 46, 242.
Burne, Owen T., schoolfellow, 45, 51.
Business interests and worries, viii, 177, 183, 186, 193, 262, 264

Caird Coast, discovered, 207.
— Sir James, 197, 198, 203, 271.
Cairn erected at Grytviken, 282.
Calgary, lecture at, 177
Campbell-Bannerman, Sir H., 95
Camping on Barrier, 67, 135 ; on glacier, 140 , on ice floes, 211–220
Canada : High Commissioner of, 269 ; visits to, 174, 176, 177, 270.
Candide, referred to, 290
Cape pigeons, 225
Cape Town, 52 , bees at, 54 ; Discovery at, 61 ; visit cancelled, 275
Captain, newspaper, 32
Cardiff, 34.
Cargo, stowing or working, 19, 39, 47, 61, 110, 113, 201, 273.
Carisbrooke Castle, serves on, 55.
Carson, Sir Edward (Lord Carson), 251.
Cary, Miss Henrietta Frances, 10.
— Rev. Henry Francis, 10
Cat on Quest, 273
Cathedral . thanksgiving service, 156 ; funeral service, 280.
Caution, 281 ; Yorkshire, 154, 290 ; of Capt England, 122
Celtic strain, 9, 99
Censorship, dislike of, 234, 253.
Cervantes referred to, 204.
Chairmen at lectures, 168.
Challenger, H.M S , 3, 16, 82 ; Reports of, 61, 82, 272.

Chanties and hymns, 291.
Character, 3, 43, 218, 287, 288.
Charcot, Dr Jean B., 105.
Charing Cross, reception at, 159.
Charities, help to, 112, 158, 163, 165, 166, 172, 187, 240, 251, 265
Charles II. and Margaret Fell, 7.
Cheeriness, 226, 288.
Cheetham, A., Endurance, 199, 205
Chest, old sea-, 10, 29, 42.
Chicago, lectures at, 175, 249.
Children : love of, 80, 91, 104, 158, 288 ; lectures to, 90, 266.
Chile, visits to, 32–33, 36, 41, 48, 49, 232, 238, 254.
China, visits to, 46, 48.
Chinaman, pony, 136, 147
Chinese labour, 95.
Chittagong, 38
Choate, Joseph, 174.
Christ Church, Westminster, 87.
Christchurch, N Z , 80, 112, 156.
Christiania, reception at, 166.
Christmas : on Barrier, 77 ; on Plateau, 142 ; on Endurance, 206 ; in Weddell Sea, 214 ; on Dublin, 260 ; at home, 186, 265, 270 ; on Quest, 277 ; lecture, 90.
Christy, Gerald, 167 ; appreciation by, 180.
Churchill, Right Hon W. S., 202.
Cigarettes, 26, 92, 148, 188, 236.
Clarence Island, 218.
Clark, R. S , on Endurance, 205.
Clive, Lord, referred to, 17.
Clyde Yacht Club, Royal, 201.
Coal . "gold" in, 25 ; at Cape Royds, 122; discovery of, 141
Coats Land, 88 , sighted, 206.
Cobden, verses on, 95.
Coffee-stall kindnesses, 288.
Commission as Major, 256, 262.
Committees . dislike of, 108, 241 ; Relief, 230, 238, 245
Comrades, tribute to, 162, 266, 280.
Comradeship, 100.
Concepcion, Chile, 41.
Confidence, 219, 235
Congo medal, 167
Congress, International Geographical, 48.
Consideration, 219.
Contemporary Review quoted, 247.
Continent, visits to the, 93, 165–171, 179, 184, 185, 269.

INDEX

Conventionality defied, 286.
Conversazione of R G S , 81.
Cook, Capt James, 7, 11, 153, 176.
— Dr F A , 164, 166, 176
Copenhagen, lecture at, 165
Cork, lecture at, 168.
Coronel, Chile, 48, 49
Courage, 10, 246, 281, 290
Cowes Week, 60, 110, 163, 200, 202.
Cowley, Abraham, referred to, 7
Craig-Lippincott, Mr and Mrs , 175
Crater Hill, 70, 73, 133.
Crean, Thomas on *Endurance*, 199, 205 , in boats, 221, 223 , crosses South Georgia, 227, 231
Credit, equipping ship on, 104.
Crevasses : on Barrier, 77, 135, 136 ; on Beardmore Glacier, 140, 146 ; on Plateau, 141
Cricket at Dulwich, 25, 37
Cromarty Firth, 260.
Cromwell and Margaret Fell, 7
Croydon, residence at South, 22
Crozier, Cape, 15, 64, 128
Crozier, Capt F. M , R N , 14
Crystal Palace, 22, 27 ; fireworks, 162
Cullum Gold Medal, 173, 294
Currie, Sir Donald, 163
C V O conferred, 162, 293.

Daily Chronicle, 197
— *Mail*, letter to, 54.
Dannebrog, Order of, 166, 293
Darkness, polar, 69, 130
Darrington, Shackletons of, 4
Darwin, Charles, 17, 55, 75, 76, 238
— Major Leonard, 159, 161
David, Prof Sir T W Edgeworth on *Nimrod*, 112, 117, 123 , at Cape Royds, 126, 128 , climbs Mt Erebus, 129, at Magnetic Pole, 134, 153, 167
Davis, Colonel Alexander, 105
— Captain John King : on *Nimrod*, 111, 157, 268 , raising funds, 185, 190 , on Ross Sea relief, 238, 241, 242
Day, Bernard C , 125, 162
Dayton, Ohio, reception at, 175
Dead man on *Hoghton Tower*, 40
Death · race with, 145–150 , wrestling with, 228 ; and knowledge, 191 , of friends, 35, 189, 197, 233, 269.

Debates on *Discovery*, 70.
Debts, 163, 188, 241, 251, 264, 267.
Deception Island, 216, 218, 220.
Decisions, quick, 118, 119, 143, 154, 221.
Defensive attitude, 286
Degree, honorary, Glasgow, 199
Dellbridge Islands, 127
Denmark, King and Queen of, 165.
Depot : in polar travel, 132 ; journeys, 128, 133, 134, 242 ; at Minna Bluff, 73, 135, 148, 149, 243 , A, 134, 136, 148 ; B, 136, 147 , C, 138, 146, 147 ; Lower Glacier, 139, 145 ; Glacier, 141 ; Plateau, 144.
Dernburg, Herr, 170
Deutschland, drift of, 189, 209, 216, 236, 247
Diary, quotations from *Discovery* expedition, 62–75 ; *Endurance* expedition, 215 ; on *Quest*, 277.
Dictating *Heart of Antarctic*, 158 , *South*, 245.
Disappointments, 55, 79, 82, 96, 118, 144, 175, 211, 263, 270.
Disaster averted, 240, 243
Discipline, speech on, 256
Discovery (1901–4), viii, 56, 58, 59, 60, 65, 79, 89, 117, 205, 280 ; (1916), 230, 232, 234.
Discovery, Mount, 148
Distinctions, list of, 293.
Docks adventure, 26 ; work in, 52
Dogs on *Discovery* expedition, 72, 76 ; on *Nimrod* expedition, 109, 129, 149 ; on *Endurance* expedition, 196, 200, 208, 215, 218, 243
Donald, Robert, help from, 197
Donne quoted, 3
Dorman, Charles (father of Lady Shackleton), 50, 57
— C Herbert, 105, 233
— Miss Emily Mary (*see also* Shackleton, Lady), 43, 49, 50, 57, 86, 87.
— Miss Julia, 93.
— Miss Maude Isabel (Daisy), 233.
Dornoch, golf at, 89
Dover Pier, wife on, 111, 112, 158
Doyle, Sir Arthur Conan, 160.
Drake, Sir Francis, 238, 285.
Dramatic instinct, 22, 178
Dreams of food, 77, 142, 148
Dresden, lecture at, 171

INDEX

Drift: of *Aurora*, 229, 243; of *Endurance* and *Deutschland*, 209, 236; of floes in Weddell Sea, 208, 215, 221
Drygalski, Dr E. von, on *Gauss*, 59
Dublin, 5, 19, 168.
Dublin, H.M.S., dinner on, 260.
Dudley Docker (boat), 219, 221, 222, 224.
Dulwich College, 24, 28, 37, 162, 228, 269
Dundee *Discovery* built at, 55, 59, election, 90–96, 267; visits to, 59, 83, 85, 90, 92, 96, 266; whalers, 39.
Dunkirk, visit to, 42
Dunlop, H J L, on *Nimrod*, 122
Dunsmore, Jas., recollections of, 50.
D'Urville, Admiral J. Dumont, 14
Duty *versus* promise, 120
Dysentery, 36, 146, 147.

Eastbourne, 55, 202, 250, 263, 265
East India Dock, 51, 59, 89
Edinburgh· residence in, 85–108; Society, 87; visits to, 168, 257, 266.
Edmonton, lecture at, 177.
Edward VII, King: on *Discovery*, 60; on *Nimrod*, 110; congratulations from, 157, confers decorations, 110, 162, 169; kindness of, 165; death of, 176.
Edwards, Thomas, 28, 125.
Effigy in Madame Tussaud's, 163.
Election at Dundee, 92–96
Elephant Island discovery of, 12, 13; landing on, 222; relief efforts, 229–237
Elizabethan qualities, 281, 285
Eliza Scott, schooner, 13
Ella of Spitsbergen expedition, 256
Emma, schooner, 233–235.
Emperor· German, 170, of Russia, 171.
Empire, love of the British, 262
Enderby Brothers, 12, 13
Enderby Quadrant, 12, 13, 271.
Endurance, 43, 146, 246, 280, 290.
Endurance· naming of, 196; equipment of, 200, voyage out, 203–206; besetment, 207; drift of, 209, 236; loss of, 210–214; film of, 264, 265; reference to, 273, 274
Enemies and friends, 286

England, Capt R. G, on *Nimrod*, 111, 116, 118, 119, 121–123.
Ensign· Blue, 52, 59, 201, 242; White, 46, 272.
Entente Cordiale, 168.
Equipment: mental, 98; of *Nimrod*, 109; of *Endurance*, 197; of Russian expedition, 257; of *Quest*, 272.
Erebus, Mount, 15, 64, 80, 121, 148, 242; ascent of, 128.
Erebus, H.M.S., 14, 15, 82.
Evans, Cape, 242, 244
— Capt E. R. G. R., R N., lecture by, 191
— Capt. F. P., on *Koonya*, 114, 116; on *Nimrod*, 149.
— Capt W A, of *Flintshire*, 51
— Petty Officer Edgar, 189.
Everett, P W, appreciation by, 84.
Eversen, Capt., of *Hertha*, 42.
Ewart, Nanty, referred to, 273.
Expedition: Arctic projected, 268; National Antarctic, 57–80; British Antarctic, 103–155; Imperial Trans - Antarctic, 193–247; Shackleton-Rowett, 268–282.
Exploration, early interest in, 24; romance of, 288.
Extravagance, 164

Failure and success, 35, 247, 289.
Fairweather, Lieut -Com, R.N., on *Discovery*, 234
Falkland Islands, 223; visits to, 231, 232, 234.
Falklands, Battle of, 205.
Falmouth, 10, 33
False Point, 39
Farthest South, Scott's (1903), 78, 137; on *Nimrod* expedition, 144; of *Endurance*, 208.
Fear, 18, 219, 221
Fell, Margaret (ancestor), 7.
Fernandino, Florida, 47.
Ferrar, Hartley T., on *Discovery*, 58, 68.
Fertilizers, scheme for making, 263
Fever, 39, 42.
"Fight the good fight," 68, 157, 291
"Fighting Shackleton," 4, 23.
Filchner, Lieut. Wilhelm, 184, 185, 189, 194, 201, 209
Film, kinematograph, 157, 161, 176, of *Endurance*, 237, 264–266; War, 254.

300   INDEX

Finland sailors from, 233; Germans in, 258.
Finse, glacier, 198.
Fir Lodge School, 23, 48, 62.
Fitzmaurice ancestors, 10, 36
— John, 9
Flag: King George's, 202, 249; Queen Alexandra's, 110, 144, 201, sledge, 67, 144 *See also* Ensign *and* Union Jack
Flattery, power of, 288
Fletcher, Phineas, quoted, 229
Flintshire: second mate on, 48, voyages, 49–51, 111.
Floes, drifting, in Weddell Sea, 206, 212, 215, 218, 219
Flower shows, 33, 187
Foca I, purchase of, 269
Food: for sledging, 132; variety of, 217; dreams of, 77, 142
Football, 25, 66
Footprints, following, on Plateau, 145.
Foreign explorers, 104, 194.
— Office, 251, 256
Fortitudine Vincimus, 8, 43, 67, 152, 196
Fortune, hopes of, 97, 164, 183, 198, 262.
Fox, George, 4, 7,
Fram, 184, 232.
Frankfurt, lecture at, 171.
Freedom, love of, 288.
French, proficiency in, 25, 169
Friends, Society of, 4, 7, 8, 9
Friendship, 117, 197, 247, 262, 287
Frobisher, Rev Anthony, 4
— Sir Martin, 4, 285.
Frost-bite, 143, 144, 220, 221, 260.
Fry, A M, 105
Fuller, Elizabeth (ancestor), 6.
Funds, difficulty as to, 195, 197, 238, 239, 269
Funeral processions and services, 278, 279, 280.
Funerals, attraction of, 19

Gaika, third officer on, 55
Galileo's birthday, 46, 66
Gambling instinct, 217.
Gardens, the Shackletons' love of, 7, 8, 9, 18, 22, 37, 289
Gardiner, Allen, 238
Gas-engines, study of, 97
Gauss expedition, 59, 85
Gavan, Miss Henrietta L. S. (Mrs Henry Shackleton), 9.

Gavan, Mrs H J., 9, 10, 19.
— Henry John (grandfather), 6, 10, 29
— Rev John (ancestor), 9.
Generosity, reckless, 84, 99, 146, 164, 289.
Geographic Society, National, 173.
Geographical: Congress, International, 48; Club, 161
Geographical Journal, 105, 111, 281.
Geographical Society: American (New York), 173, Austrian, 170, 193; Danish, 166; German, 170; Hungarian, 170, Italian, 158; Royal (London), *see* Royal; Royal Scottish, *see* Scottish; Russian, 171; of Paris, 168; of Philadelphia, 173.
Geological specimens, 140, 151.
George v, King on *Nimrod* as Prince of Wales, 110; on *Endurance*, 200, 202; audiences of, 200, 249, 260, 273; congratulations from, 231; lecture to, 252
Gerlache, Capt Adrien de, 167.
German Antarctic expeditions, 59, 184, 189, Army in Finland, 258, Emperor, 170; fleet, 205; language, 25, 170, 179, 185; influence in S America, 253, 254; spies, 255.
Germany, visits to, 169, 179, 185.
Germs in Antarctic, 200
Glacier Bay on Caird Coast, 207
— Beardmore, *see* Beardmore; Tongue, 73, 127, 133.
Glasgow lecture at, 266; University degree from, 199; work in, 97, 103.
Goggles, for snow, 139
Golden Hind, Drake's, 238, 285
Golf first drive, 86; at Buenos Aires, 254; at Dornoch, 89; at Seaford, 188; at St Andrews, 92.
Good Hope, storm off Cape of, 38
Gothenburg, visit to, 166
Government grant for *Nimrod*, 164; for *Endurance*, 198, for relief expeditions, 238.
Graham Land, 13, 42, map, 236.
Gran, Tryggve, 167.
Granite Harbour, 64.
Gratitude, example of, 146.
Gratz, lecture at, 179

INDEX

Gravedigger, early ambition, 21.
Green, C. G. (cook), 274.
Green's shipyard, 109.
Greenstreet, L., on *Endurance*, 205.
Grey, Earl and Countess, 174.
Griese, river, 18.
Griffiths, Lieut.-Com. Arthur, R.N.R. (Griff), 26, 260.
Grisi, pony, 138, 147.
Grytviken : *Endurance* at, 204, 205; *Quest* at, 277; burial at, 279; monument at, 282.
Guarantors, 103, 104, 164, 264.
Guildhall, lectures in, 187.
Guthrie, Mrs. Hope, appreciation by, 98.

Half-Moon Bay, 155.
Halifax, lecture at, 168.
Halsbury, Earl of, 160.
Hamburg, 33, 47; lecture at, 170.
Harbord, E. A., on *Nimrod*, 111.
Harbour Hill, 73.
Harness, sledge, 140.
Harrow School, visits to, 266.
Harvard, enthusiasm at, 173.
Harwich Yacht Club, Royal, 59.
Hastings, lecture at, 172.
Hayward, V., on *Aurora* expedition, 242, 243, 244.
Heart of the Antarctic: writing of, 158, 163; publication of, 167; French translation, 168; quoted, 137, 142, 143, 144, 146, 155.
Heat: dislike of, 254, 275; in the Antarctic, 138, 215.
Hecklers: at Dundee, 95; at Philharmonic Hall, 267.
Hedin, Dr. Sven, 166.
Hegedüs, Sandor von, 186.
Heinemann, William, 158, 160.
Henry of Prussia, Prince, 171.
Hero Readers, 187.
Hero-worship, 190.
Higgins, Miss, Fir Lodge School, 23.
Hinton Charterhouse, 197.
Hobbs, Prof. W. H., notes by, 175.
Hodgson, T. V. ("Muggins"), on *Discovery*, 58, 70, 107.
Hoghton Tower: first voyage on, to Chile, 30, 239, 275; second voyage to Chile, 34; third voyage to India, Australia, and Chile, 37; aground, 39; dismasted, 41, 42; as simile, 161.

Home-coming, 33, 36, 49, 51, 81, 158–160, 249, 255, 259, 263.
Home Rule for Ireland, 19, 95.
Home-sickness, 30, 37, 112, 288.
Honours, ambition for, 22; appreciation of, 156, 289; list of, 293–294.
Hooker, Sir Joseph Dalton, 14, 82.
Hope, Mount, discovery of, 138, 139; depot at, 243.
Hopkins, Capt. J. B., 30.
Horn, Cape, storms off, 32, 36, 233.
Horoscope from ship, 3.
Horse, holding a, 177, 287.
Horse-flesh, 138, 146, 147.
Hospitals, helping, 158, 165, 177, 265.
Hovgaard, Commander, 166.
House: Kilkea, 18; in Dublin, 19; in Sydenham, 22; in Edinburgh, 87; at Putney Heath, 185; at Vicarage Gate, 195; Eastbourne, 250.
Hubbard Gold Medal, 173, 294.
Hudson, H., on *Endurance*, 205.
Huguenot ancestors, 10, 169.
Humour, 178, 217; American, 176.
Hungary, visits to, 170, 184, 186.
Hunger, 77, 138, 142, 148, 288.
Hurley, F., on *Endurance*, 205.
Hussey, Capt. John Austen, appreciation by, 56.
— L. D. A.: on *Endurance*, 205; on Elephant Island, 237; at home, 247, 257, 266; at Murmansk, 258; on the *Quest*, 274; in Montevideo, 278; at Grytviken, 279; appreciation by, 280.
Husvik, South Georgia, 229.
Hut: at Aberdeen House, 25; at Cape Royds, 125, 126, 151, 242; at Hut Point in 1902, 66; in 1908, 133; in 1909, 151; in 1916, 243; at Cape Evans, 242, 244.
Hut Point, 73, 127, 133, 151, 243.
Hutchison, Alfred, 200, 231.
Huxley's *Lay Sermons*, 74.
Hymns: on the *Hoghton Tower*, 31; favourite, 68, 157, 291.

Ice Barrier: off Coats Land, 206; the Great, *see* Barrier.
Icebergs, early picture, 21; in Ross Sea, 65, 115; off South Georgia, 277; in Weddell Sea, 206; risks from, 188.

Ice-falls on Glacier, 141, 146
Ideals, 281, 289
Ignis fatuus, 93, 188
Imagination, 99, 261, 281
Imperial Trans-Antarctic Expedition, 193–247, route, 196
Inaccessible Island, 127, 242
Inconsistency, 287
India, passages on, 112, 158
Indian Ocean, 38
Influenza at Grytviken, 204, at St Vincent, 275
Information, Department of, 251.
Insight, quickness of, 218
Instituto de Pesca, 232, 279
Intuition, 154
Invasion, winter, plan of, 250
Invergordon, sails from, 260
Invernairn, Lord *See* Beardmore
Iquique, visits to, 33, 36
Ireland, lectures in, 168
Irish . ancestors, 9 ; brogue, 23, intelligence, 154, wit, 290
Irizar, Capt Julian, 83
Ironside, Maj - Gen Sir W G, 258.
Isis, passage on, 158
Islands, doubtful, 157, 268
Italy, King of, reception by, 169

James, R W , physicist on *Endurance*, 205, 216, appreciation by, 217–218
James Caird, boat : in Weddell Sea, 213, 219, 221–223 ; voyage to South Georgia, 224–227 ; at Albert Hall, 205, presented to Dulwich College, 228.
Jane, brig, 12
Japan, voyages to, 46, 48, 49
Jockey Club, Santiago, 239.
Johnson, coachman, 46.
— Dr Samuel, 6, 7
Joinville Island, 218
Jokes, practical, 22, 92, 178, 269
Journalist, life as, 84
Joyce, Ernest, on *Nimrod*, 108, 125 ; laying Bluff depot, 135, 149 ; receives medal, 162 ; on *Aurora*, 242
Jumping Jenny, referred to, 273

Kaiser, lecture to the, 170
Kaiser Wilhelm Land, 85
Kay, Chris , schoolfellow, 26.
Keats, poetry of, 56.

Keedick, Lee, 175, appreciation by, 180
Keighley, Shackletons of, 4.
Kerr, A , on *Endurance*, 205
Kiel, lecture at, 171.
Kilkea, Co. Kildare, 3, 17, 19, 280
Killarney, private railway car, 177.
Killer-whales, 219, 220.
Kindness, 43, 288
Kinema films, 157, 254, 265.
Kinematograph, first in Antarctic, 106, 162.
King Edward Land discovery, 65, projected base at, 111, 184, attempt to reach, 118–120, 207
King Haakon Fjord, 226.
Kinsey, Sir Joseph J , 122.
Kipling, Rudyard, meeting with, 54 ; poetry of, 99, 290 ; quoted, 45, 81.
Kitchener, Earl, 174, 203.
Knighthood conferred, 168, 293.
Koettlitz, Dr Reginald, on *Discovery*, 58, 63, discovers moss, 64
König, Dr Felix, projected expedition, 193, 197, 201, 202.
Königsberg, lecture at, 179.
Koonya tows *Nimrod*, 114, 115
Kosmos Club, 105.
Kristensen, Capt Leonard, 46.

Labour Members, value of, 255
" Lag, Mr ," early nickname, 18
Lambton, Miss Elizabeth Dawson, 104, 198.
Landing by boat at Elephant Island, 222, 236 ; on South Georgia, 226
Lands of Silence, error in, 79.
Lanternists, incidents with, 83, 175, 176.
Lapland, passage on, 252.
Larsen, Capt. C. A., 42.
Latin, Abraham Shackleton and, 5, 170
Lauder, Sir Harry, 192.
Laurier, Sir Wilfrid, 174.
" Lead kindly light," 157, 291.
Leadbeater, Mary, 6
Leadership, gift of, 122, 153, 160, 219, 281
Leap Year Day, 1916, 216.
Lecture agents, 111, 167, 174, 175, 178, 180

INDEX

Lectures . in Albert Hall, 161, 265, in Philharmonic Hall, 266–268; at Southport, 83; to children, 90, 266; on Continent, 165, 169, 171, 179, 184; in America, 173–177; in Great Britain, 168, 172, 178, 187, 192, 199, 263, 266, on board liners, 112, 172, 253; at Rio, 276.
Leeds, lecture in, 163.
Legion of Honour, 169, 293
Leipzig, lecture at, 171.
Leopold II, King, 93, 167.
Letter · from Mr. Asquith, 164; from Sir James Caird, 198; from Sir E. Carson, 251; from King Edward, 157, from Dr E A Wilson, 106; anonymous, 254; to a friend, 103, 104, 186, 241, 255, 270; to parents, 31, 35, 40; to daughter, 233; to son Edward, 235; to son Ray, 224, 258; to his wife, 50, 112, 118–121, 165, 203, 204, 240, 246, 252; to J Q. Rowett, 276; to Dame Janet Stancomb-Wills, 276; to J Sarolea, 250
Letters, love of receiving, 38, 42, 155
Lewisham, reception at, 162
Liberal Unionist Party, 90, 189.
Lifeboat Fund, lecture for, 163
Life-saving, 220
Lion of the season, 159
Lips, blistered, 148.
Lisbon, *Quest* at, 275.
Lister · Lord, 199; Mount, 199.
Lively, cutter, 13
Liverpool, 30, 33, 37, 203, 252
Livingstone Medal, 168
Lloyd's, Committee of, help, ix
Loan in New Zealand, 241, 244
Lóczy, Dr. Lajos, 170.
London Season, 1909, 159
Longfellow, poetry of, 35
Longhurst, Cyril, 87
Longstaff: Ll. W., 58, 271, Mount, 78, 137.
Lowestoft, lecture at, 172.
Loyalty, 261, 280.
Lucas, St John, quoted, 265
Luck, good, 153, 224
Lucy, Sir Henry, help from, 163
Luitpold Land, 189, 208.

Lusitania, lecture on, 172.
Lysaght, Gerald S.: meeting, 53; on *Quest*, 274.
— Sydney, 105, 112.
Lyttelton, N Z.: *Discovery* at, 61, 62; *Nimrod* at, 113, 155.

M'Carthy, T., on *James Caird*, 223.
M'Clintock, Admiral Sir Leopold, 10.
MacDonald, Alan, 253.
McIlroy, Dr. J A · on *Endurance*, 205; on Elephant Island, 237, 247; in Spitsbergen, 256; on *Quest*, 268, 274, 278.
Mackay, Dr. A Forbes, on *Nimrod* expedition, 108, 125, 129, 134, 151
Mackellar, Campbell, 105
Mackintosh, Capt Æneas L A., on *Nimrod*, 111, 151; accident to, 121; gets medal, 162; on *Aurora*, 199; on the Barrier, 230, 243; at Cape Evans, 242; death of, 244
— Mrs, lectures for, 245
Macklin, Dr. A H., on *Endurance*, 205, on Elephant Island, 237, 247; at Murmansk, 258; on *Quest*, 274, 278.
M'Lean, Dr. W., of *Tintagel Castle*, 53, 55
M'Murdo Bay, discovery of, 15
— Sound: *Discovery* in, 66; description of, 127; as base for *Nimrod* expedition, 106, 118; in *Aurora* expedition, 243–244
M'Nab, Dr Robert, 241.
M'Neish, W., on *James Caird*, 223
Madeira: *Discovery* at, 61; *Quest* at, 275
Madras, call at, 38
Magellan, Ferdinand, 238
— Strait, 232
Magnetic pole, David's discovery, 134, 150, 157; research, 14
Maize as sledging food, 140.
Major, commission as, 256
Manchuria, ponies from, 109.
M A P quoted, 24
Map Propagandist, 254, of British National Antarctic Expedition, 124, of projected Trans-Antarctic expedition, 196; of Antarctic regions, 1874, 11, of *Endurance* expedition, 236.

804 INDEX

Margate, *Endurance* at, 202.
Markham: Sir Clements, 58, 67, 72, 79, 81, 89, 160, Mount, 78, 137.
Marlborough Club, election to, 165.
Marriage to Miss Dorman, 87
Mars, H.M.S., 260.
Marshall, Dr Eric S., on *Nimrod*, 108, 126; climbs Mount Erebus, 129; depot-laying, 134, 135, on Southern journey, 141, 144, 149, gets medal, 162, at Archangel, 263
Marston, George E, on *Nimrod*, 108, 125, 134, gets medal, 162; book by, 191; on *Endurance*, 205.
Masefield's poetry, 290.
Master's certificate, 49.
Mate, examination for, 43, 48
Mathematician, simile of, 290
Mathew, Father, 8
Mauretania, passage on, 190
Mauritius: visit to, 39, sailor from, 233
Mawson, Sir Douglas on *Nimrod*, 112, 126, climbs Mount Erebus, 129; at South Magnetic Pole, 134, 150, 153; on Australian Antarctic expedition, 185, 190, 195, on Relief Committee, 231
Maynard, Maj.-Gen Sir C. M., 258, 259; appreciation by, 261
Medals, awards of, 89, 158, 162, 166, 168, 169, 173, 294.
Mediterranean, first voyage on, 46.
Melbourne, lectures at, 112, 158
Melbourne, Mount, 63
Mellor, Lydia (ancestor), 7
Memorial: at Grytviken, 282; service in St Paul's, 280
Mercantile Marine, career in, 30–55.
Mermaid Tavern referred to, 286.
Meteorological data, 70, 246
Micawber, Mr, referred to, 290.
Michigan, University of, 175
Micky or Mike, nickname, 23, 260
Middlesex Hospital, help to, 265.
Milk, dried, 110
Mill, Dr H R., on *Discovery*, 59
Mills, Sir James, 114
Milton's poetry, 7, 35, 290
Minna Bluff, 73, 148
Mirage reveals depot, 149.
Mirni, Russian ship, 12.

Mischief, love of, 17, 286.
Mobile columns, 259.
Modesty, 156, 187.
Monmouthshire, voyages on, 45, 48.
Montelius, Professor, 166.
Montevideo, visit to, 240; funeral at, 278, 279.
Montreal, visit to, 177.
Monument, The, 22
Moone, Co Kildare, 8, 17, 42.
Moore, Lieut T E. L., R N, 15.
Moraines, Beardmore Glacier, 140.
Morley, Viscount, quoted, 6.
Morning: return on, 79; relief ship, 82, 89
Morrell Benjamin, 12, 247; Land, 13, 189, 209, map, 236.
Moscow Soviet, 258.
Moss discovered in Antarctic, 64.
Motor-car, first in Antarctic, 106, 121, 133, -crawler and warper, 208; -sledge, 196.
Mountaineering, 20, 73, 227
Mountains: on Beardmore Glacier, 139, of South Georgia, 204, 227, 279.
Mulock, Lieut. G. F. A., 79, 107.
Munich, lecture at, 171.
Murmansk, stay at, 257, 260, 263.
Murray, George, F.R S., on *Discovery*, 59.
— James, biologist: on *Nimrod*, 108, 125, at Cape Royds, 135; gets medal, 162; writes *Antarctic Days*, 191.
— Sir John: on *Challenger*, 16, 82; in Edinburgh, 86, 89, 91; lake survey, 108.
M V O, conferred on board *Nimrod*, 110.

Nagasaki, buying books at, 48
Nanoose, private car on C P R., 177.
Nansen, Dr. Fridtjof, 67, 108, 160, 166, 232
Nares, Admiral Sir George, 16
National Gallery, visits to, 49.
Nature, love of, 44
Naval Reserve: joining, 55, 58, resigning, 82
Navarro, Antonio de, appreciation by, 172
Neptune: on *Hoghton Tower*, 32; on *Tintagel Castle*, 53
Newbolt, Sir Henry, quoted, 248
Newcastle, N S.W., visit to, 39, 40.
New England, lectures in, 173.

INDEX

Newnes, Sir George, 271
New Orleans, visits to, 240, 255.
Newport News, visit to, 47.
News Agency, projected, 85.
Newspapers, arrangements with, 111, 154
News service as propaganda, 253
New South Greenland, 13, 247
New Year's Day, 1893, at Chittagong, 39; 1894, at Valparaiso, 42; 1896, in Red Sea, 48; 1899, at home, 51; 1902, on *Discovery*, 63; on *Nimrod*, 114; 1909, on the Plateau, 143, 1910, in Rome, 169; 1922, on *Quest*, 277.
New York, visits to, 47, 48, 80, 173, 174, 189, 240, 249, 252
New Zealand: on *Discovery*, 61; on *Morning*, 80; on *Nimrod*, 113, 114, 154–157; on *Aurora*, 239, 241
—— Government, kindness of, 112, 114, 238, 241
Nicol, Sir W. R., tribute by, 160.
Nimrod: bought, 109, leaves England, 110; towed to Antarctic Circle, 114; making for King Edward Land, 118, 203, 207; at Cape Royds, 121, 123; at Hut Point, 151; at Stewart Island, 154; arrives at Lyttleton, 155; voyage home, 157; at Temple Pier, 165.
Nordenskjöld, Dr Otto: in Antarctic, 60, 83; at Gothenburg, 166.
North, Cape, *Nimrod* off, 152
Northcliffe, Viscount, 271.
Northern Exploration Co , 256
North Russian expedition, 256–262
North Sea, 51, 263
North-Western Shipping Co , 30, 34
Norway, visits to, 108, 166, 198, 256, 269
Norwegians in Antarctic, 42, 46, kindness of, in South Georgia, 227–229, 279.

Oates, Capt L E G , death of, 190
Oban, N Z , 155
Observations for position, 74, 144, 216, 225
Obstinacy of character, 33, 43.
Ocean Camp, 214, 216.

Oceanographical expedition, project, 268
O'Connell, Daniel, 8.
O H.M.S., Shackleton's first book, 53, 55
" Old Cautious," nickname, 221.
Omar Khayyam, 63
Omnibus system of London, 263.
Optimism, vii, 99, 132, 191, 199, 203, 213, 214, 223, 224, 246, 263, 290.
Orde-Lees, T., on *Endurance*, 205
Orders, list of British and Foreign, 293
Organization, power of, 21, 99, 153
Orion, 192.
Orotava, passage on, 80
Osternieth, Madame, 167.
Otira gorge, visit to, 80
Ottawa, lecture at, 174

Pacific Ocean, storm in, 40, 41.
Pack-ice in Ross Sea, 63 ; at Bay of Whales, 118 ; in Weddell Sea, 205; near Elephant Island, 231, 232, 233
Packing-cases, new design, 110.
Pagoda, barque, 15.
Panama, travels by, 240, 255
Pardo, Capt Luis, on *Yelcho*, 235
Paris, lectures in, 168
Parkhead Engineering Works, 97.
Parliament : candidate for, 92–97 ; grant by, for *Nimrod*, 164 ; requests to stand for, 90, 189.
Partridge, Captain, of *Hoghton Tower*, 30, 33.
Patience, 128, 215
— Camp, 215–220.
Patriotism, 248
Paulet Island, 211, 214, 218.
Pearson, Sir Arthur, 84.
Pearson's Magazine, 84.
Peary, Rear-Admiral R E , U S N , 164, 172, 173, 232
Peggotty Camp, 227, 231.
Penck, Prof. Albrecht, 170
Penguin . muff, 18, 19 ; toy, 269
Penguins first seen, 63 ; for food, 127, 206, 224.
Perseverance, virtue of, 290
Persia, East Indiaman, 10
Personality, 100, 280, 290
Personnel : of *Discovery*, 58 ; of *Nimrod*, 108, 125 ; of *Endurance*, 199, 205 ; of *Quest*, 274
Persuasion, powers of, 21, 204

INDEX

Perth, W. A., lecture at, 158.
Pessimism, guarded against, 217.
Peterborough, honeymoon at, 87.
Philadelphia, visits, 173, 174, 249
Philharmonic Hall, lectures, 266
Phillips, Stephen, poetry of, 63.
Phonograph, first in Antarctic, 106
Photographs, 63, 157, 240.
Piper, Highland, at *Quest's* send-off, 202
Pittsburgh, lecture at, 249
Plans: for *Nimrod* expedition, 105, 106–111; for *Endurance* expedition, 194; for *Quest* expedition, 270, change of, 118, 121, 204, 213, 276
Plasmon biscuits, 110
Plateau, South Polar, reached, 142, annexed, 145; farthest camp on, 170
Pluck and pertinacity, 238
Plum pudding for Antarctic Christmas, 77, 142
Plymouth, lecture, 172; *Endurance* leaves, 203; *Quest* leaves, 273
Poetry, love of, viii, 7, 10, 21, 35, 52, 62, 98, 130, 217, 281, 287
Polar (or Arctic) Medal, award of, 89, 169, 293
Polaris, renamed *Endurance*, 196
Pole, South, problem of reaching, 131, reached by Amundsen, 186, by Scott, 189
Poles in Lanarkshire, 95
Poleward progress, table of, 153
Ponies in Antarctic, first, 106, 115, 121, 129, 134, 139
Popularity, 156–180, 251.
Port Chalmers, sails from, 62, 242
Port Louis, Mauritius, visit to, 39
Port Stanley, visits, 231, 232, 233
Portland, Oregon, visit to, 49; lecture at, 249
Posen, lecture at, 179
Possession Island, 15
Postmaster in Antarctic, 116
Pourquoi Pas? Charcot's ship, 105
Power, love of, 246
Prague, lecture at, 179
Pram Point, 68; revisited, 127, 150, 151
Prayers on *Hoghton Tower*, 31; on *Discovery*, 69.

Presence, not of this world, 227, 246
President, H.M.S., 58.
President: of Chile, 239, 240; of France, 169; of United States, 173, of Uruguay, 240, 279
Pressure ridges in ice-floe, 210
Pride that apes humility, 231.
Priestley, Raymond E., geologist: on *Nimrod*, 108, 125; gets medal, 162, revisits Cape Royds, 151; appreciation by, 130, 132
Prinz Luitpold Land, 189, 208
Procession at Punta Arenas, 240; at Christiania (torchlight), 166, funeral, 278, 279.
Professor Gruvel, steamer, 278.
Promise to Capt Scott, 108, 119, 157
Propaganda in S America, 251
Prospectus of lectures, 89, of 1914 expedition, 197
" Prospice " quoted, 163, 291
Providence, guidance of, 146, 149, 154, 227, 246, 291
Psychology, Polar, 117
Ptomaine poisoning, 147.
Public opinion, 204
Publicity, excessive, 271
Punta Arenas, visits, 232, 235, 237.
Putney Heath, house at, 185.

Quaker ancestors, 4–8, 22, 242
Quan, pony, dies, 138
Quarrels on expedition, 62.
Quebec, sails from, 177.
Queen See Alexandra, Queen.
Queen Mary Land, 190
Queen's Hall, lecture in, 162.
Queensferry, summer at, 97.
Queenstown, call at, 42
Quest naming of, 270; equipment of, 272; voyage of, 273–277.
Quotation: love of, 21; unverified, 191, 265, 289

Race with death on Barrier, 145
Rain in the Antarctic, 218
Raleigh, Sir Walter, resemblance to, 281, 285
Ramsgate, lecture at, 251.
Rats on *Hoghton Tower*, 40
Reading, love of, 28, 35, 46, 50, 71.

INDEX

Reception: in Berlin, 169; in Copenhagen, 166; at Charing Cross, 159; in Christiania, 166, in Dayton, Ohio, 175, in Lyttleton, 155; in Punta Arenas, 237, 238; in San Francisco, 249; in Santiago, 239; in South Georgia, 227; in Valparaiso, 239, in Wellington, 244.
Recruiting address, 248–249.
Redcar, *Flintshire* ashore at, 51.
Red Cross Society, 240.
Redgauntlet, referred to, 273
Red Sea, 46, 48
Regina, lecture at, 177.
Relaying sledges, 75, 141
Relief expeditions: for *Discovery*, 82, 83; for Elephant Island, 230–237, for Ross Sea party, 241–244.
Religious views, 28, 35, 43, 291
Reporters, manner to, 157, 158, 180.
Reports, false, 79, 123.
Research, method of helping, 200.
Resolution, H M S, 7.
Responsibility, sense of, 117, 203, 214, 224, 230.
Rest, desire for, 233, 246, 263, 277.
Restlessness, 49, 90, 103, 192
Results: of *Nimrod* expedition, 152; of *Endurance* expedition, 246
Rhinoceros hunt, 38.
Rhodes, S. G., schoolfellow, 23, 48
Rich men, search for, 104, 109, 194, 197, 268, 269
Richards, W. R, on *Aurora*, 242.
Rickinson, L, on *Endurance*, 205
Rights relinquished, 241.
Ringaroona, H M S, 62.
Rio de Janeiro, visit to, 253; stay at, 275–276.
Rio Secco, 237.
Risks: reasonable, 18, 211, 223, unreasonable, 20, 133
Roberts, William C, cook, 125, 162
Robertson, Edmund, M.P, 95.
— Capt. Thomas, of *Scotia*, 89.
Robinson, Capt Robert, of *Hoghton Tower*, 34, 42.
Robson, Henry, opponent at Dundee, 95
Romance: in Sydenham wood, 26; of exploration, 281, 288
Rome, lecture in, 169
Roosevelt, Arctic ship, 232
Rosebery, Earl of, 88, 165, 199, 200.

Ross Island, description of, 127.
Ross, Sir James C., 14, 46, 88, 153
Ross Sea: exploration of, 14, 46, *Discovery* in, 63; *Nimrod* in, 116; *Aurora* in, 195, 199, 230; relief expedition, 238, 239, 242
Rowett, John Q., at Dulwich College, 24; lends office, 185; finances *Quest*, 269, 271; letter to, 276
Royal Geographical Society promotes exploration, 13, 58, fellowship of, 53, conversazione of, 81; awards silver medal, 89, helps *Nimrod* expedition, 110; lectures to, 161, 188, 266; awards special gold medal, 162, helps *Endurance* expedition, 198; represented on relief committee, 230; at memorial service, 280
Royal Magazine, sub-editor of, 84
Royal Naval Reserve, 55, 58, 82
— Navy: accepts conditions of, 59; tries to join, 82
— Scottish Geographical Society See *under* Scottish
— Societies Club geographical table, 85, welcome by, 160.
Royds, Capt Charles W. R, R N, 58
— Cape: discovered, 73, selected as base, 121, hut at, 125, 151; revisited, 242.
Rugby School, lecture at, 168
Rum cart, invention of, 72, 133
Russia: Emperor of, 171; Dowager Empress of, 166, 201; visits to, 171, 257; concessions in, 97, 263.

Sabine, Mount, 63
Sabrina, cutter, 13.
Saigon, visit to, 49
Sail on sledge, 147, 148
Sailor, British merchant, 30, 43, 94, 96.
St Andrews: golf at, 92; lecture in, 172
St Andrew's Night at Rio, 275.
St Katherine's Dock, 273
St. Patrick's Day fight, 23, 48
St. Paul's: Cathedral, Memorial Service, 280; Rocks, 32, 275.
St. Petersburg, lecture at, 171.

INDEX

St. Vincent, Cape Verdes, on *Flintshire*, 48, 49, on *Quest*, 275
Sale-Barker, Mrs, 35
— Maurice, 22.
Sandringham, lecture at, 252.
San Francisco, visits to, 48, 80, 240, 249
San Sebastian, Argentine, 233
Santiago reception at, 239; visit to, 254
Sarolea, Professor Charles, 93, appreciation by, 247.
— Jack, letter to, 250.
Saunders, Edward, literary assistant, 158, 245
Savage Club dinner, 160
Schokalsky, General Jules de, 171
School. Preparatory, at Fir Lodge, 23, reports at Dulwich College, 24, readers, 187
Science, services to, 200, 246, 249
Scientific Staff, selection of, 200
Scotia, voyage of, 60, 89, 247
Scotland lecture tours in, 168, 178, residence in, 86–108
Scott, Capt. Robert F., R N, appointed to National Antarctic Expedition, 55, commands *Discovery*, 58; balloon ascent, 66, plans Southern journey, 71, starts, 75; turns at 82° 15′ S, 78; denies false report, 79; returns home, 89; objects to plan of British Antarctic Expedition, 107, keeps plans secret, 111; Shackleton's promise to, 107, 120; shows no resentment, 157; receives Shackleton, 159, 160, sails in *Terra Nova*, 184; reaches South Pole and dies, 189; Memorial Fund, 190, tribute to, 191
— Sir Walter, works of, 35, 88.
Scottish Geographical Society, Royal lectures to, 82, 86, 90, 168, 266; secretary of, 85; offers resignation, 91; resigns from, 92; awarded medal by, 168
Scurvy: attack of, 78; *Aurora* party attacked by, 243.
Sea-anchor, lying to, 221.
Seabirds and souls of the dead, 36.
Sea-blisters, 224.
Sea-chest, old, 10, 29, 41

Seaford, gol at, 188
Sea-ice in M'Murdo Sound, 128, 133, 242, 243.
Sea-plane, 275
Sea-sickness, 10, 31, 34, 221, 274
Seals · in Ross Sea, 64; in Weddell Sea, 206, 209.
Seamen's Hospital, Poplar, 172
Seattle, lecture at, 249.
Self-confidence, 132
Sellar, Mrs W. Y., 88.
Semenoff-Tianshansky, Dr, 171.
Sensitiveness, 56, 173, 286.
Service, Robert. appreciation of, 166, 174, 290; quoted, 125
Sextant lost, 47; use of, 74, 225
Shackleton, Abraham (born 1696), 5, 170
— Abraham (born 1752), 6, 8.
— Miss Aimée V (sister), help acknowledged, ix.
— Miss Alice (sister), help acknowledged, ix.
— Cecily Jane Swinford (daughter), birth, 103; 203; letter to, 233.
— Ebenezer (born 1784), 8.
— Edward Arthur Alexander (son), born 1911, 185, 235
— Miss Eleanor (sister), help acknowledged, ix, 177
— Miss Ethel (sister), 43
— Ernest Henry. birth, 3, 10; ancestry, 4–11, first penguin, 19; life in Dublin, 19; in Sydenham, 22; at Dulwich College, 24; playing truant, 26, as boy on *Hoghton Tower*, 27, 30, apprenticeship, 34; meets Miss Emily Dorman, 43; mate on Shire Line, 45, officer on Union Castle Line, 51, writes first book, 53; lieutenant on *Discovery*, 56, 57; engaged to Miss Dorman, 57, chosen by Scott for Southern journey, 71; at 82° 15′ S., 78; invalided home, 79; sub-editor of *Royal Magazine*, 84; Secretary of Scottish Geographical Society, 85; married, 87, stands for Parliament, 90; joins firm of Beardmore, 97; plans Antarctic expedition on new lines, 104; makes promise to Scott, 107,

INDEX

Shackleton, Ernest Henry (contd.)—
leads *Nimrod* expedition, 114; a broken promise, 119; reaches 88° 23' S, 144, rewards and lecturing tours, 156–180; knighted, 168, business enterprises, 183; plans trans-Antarctic expedition, 194; leads *Endurance* expedition, 204; retrieving failure, 213–247; rescues comrades, 237, acclaimed in S America, 238, excluded from command of Ross Sea rescue, 238, signs on under Davis, 241; recruiting speeches, 248; secret mission to S. America, 251; on North Russian expedition, 257; film lectures, 265; organizes Shackleton-Rowett expedition, 270; sails on *Quest*, 273; death at South Georgia, 277; a great funeral, 279; burial at Grytviken, 280; character and qualities, 285–291; list of distinctions, 293.

— Lady (Mrs E H) help acknowledged, vii, viii, ix, marriage, 87; in Edinburgh, 87–88, golfing, 92, 188; help for expedition, 104, on Dover Pier, 111, 158; received by King and Queen, 162; on continental tour, 166, 168, 169; on American tour, 172, 174, 175, 177; describes practical joke, 178; visit to Sir James Caird, 203; hears news of safety, 231; received by Queen Alexandra, 251; farewell dinner in Edinburgh, 257; long distance telephone talk, 260; decision as to burial, 279; letters to, 112, 118–121, 190, 203, 204, 240, 246, 252, Appendix by, 293 *See also* Dorman, Miss Emily.

— Henry (married 1588), 4
— Dr. Henry (father). birth, 8, at Kilkea, 9, 18; in Dublin, 19–22; at Sydenham, 22, love of flowers, 60; referred to, 47, 159, 178; illness, 266; death, 269

Shackleton, Mrs Henry· help acknowledged, viii; marriage, 9; ancestry, 9, 10; at Kilkea, 18; in Dublin, 18.
— John (16th century), 4
— Mary (Mrs Leadbeater), 6.
— Raymond Swinford (son), 91, 203, letter to, 258
— Richard, arms of (1600), 8
— Richard, at Flodden, 4.
— Richard (first Quaker), 4
— Richard (friend of Burke), 5, 6, 7, 242
— Roger (born 1616), 4.
— House, near Bingley, 4, 5; village, Yorkshire, 4
— Inlet, 78
Shackleton-Rowett expedition, 271
Shakespeare. love of, 290; quoted, 99.
Shaughnessy, Sir Thomas, 177.
Sheep at Antarctic Circle, 115.
Sheringham, holiday at, 177.
Shetlanders at Grytviken, 279
Shirase, Japanese explorer, 184
Shire Line, the Welsh, 45
Sidney, Sir Philip, recalled, 146
Sight-seeing in Budapest, 170, in Rome, 169.
Signalling, facility in, 53, 97.
Simon's Bay, *Discovery* at, 61
Singapore, Naval Court at, 50
Sirius, 53, 192.
Skelton, Eng.-Lieut R. W, R N, on *Discovery*, 58, 63, 79, 107
Ski-running, 63, 66
Slatin Pasha, 165
Sledge: harness, 140, meter, 145, 146; sailing, 145, 148; travel, longest, 242
Sledging. and equipment, 131, 262; autumn, 128; for children, 90; rations, 223; spring, 73, 133
Sleep, Ned, schoolfellow, 26.
Smith, A. Duncan, 93
Smoking, earliest, 26, 33.
Snow blindness, 76, 78, 136, 139, bridge, 77; pillars, use of, 138, 147.
Society: in Edinburgh, 88; in London, 160, 187; in New York, 189
Socks, pony, loss of, 140
Sorlle, Mr, in S. Georgia, 227, 229
Soroka, 263.

INDEX

Soundings, deep, in Weddell Sea, 88, 209.
South America, propaganda in, 251
South: dictation of passages, 245; publication of, 263; royalties on, 264; quotations from, 210, 211, 219
South Georgia (Isle of Georgia) discovery, 7, 11, 12; *Deutschland* at, 189; *Endurance* at, 204; *James Caird* at, 226; crossing of, 227; *Quest* at, 276, 277; funeral at, 279, 280
South Pole: Amundsen at, 166, 186, Scott at, 189; approach to, 131, 153
South Polar Times, 70, 74
South Sandwich Group, 205
— Shetlands discovery of, 12; whalers at, 42; boats make for, 216, 218
— Trinidad, visit to, 61
Southern Cross, voyage of, 52, 58
Southern Sky, 229, 231
Southport, British Association, 83
Soviet government, 258
Spencer Smith, Rev A, breakdown and death of, 243
Spies, dogged by German, 255
Spirits: good, 237, 271, 273; depressed, 38, 41, 175, 240
Spitsbergen, expedition to, 256.
Stamps, Antarctic, 116
Stancomb-Wills, Dame Janet, 198, 219, 251, 276.
Stancomb-Wills, boat, 216, 219, 221, 223, 224
Stars, mystical meaning, 44, 49, 53, 192, 277, 291
Staten Island, 233
Stefansson, Vilhjalmur, 269
Stenhouse, Capt J. R : on *Aurora*, 230; regard for, 242, 247, 268; at Murmansk, 258.
Stevens, Rev. Henry, 35
Stevenson, R L : characters of, 26; memorial, 282.
Stewart Island, 109, 154, 155
Stockholm, reception at, 166.
Stores, work on : on *Discovery*, 61; on *Morning*, 80; on *Terra Nova*, 83, on *Nimrod*, 110, at Cape Royds, 121, 126; on *Endurance*, 203, 214; at Murmansk, 259; on the *Quest*, 272.

Storms at sea · in Channel, 10, on *Hoghton Tower*, 32, 38, 40, 41, 42; in North Sea, 51, 263; on *Nimrod*, 115; in Weddell Sea, 207, in *James Caird*, 226; during Elephant Island relief, 233, 235; in M'Murdo Sound, 244; on the *Quest*, 277
Story-telling, 22, 26, 84, 98, 217, 289
Strathcona, Lord, 162.
Stromness Whaling Station, 227.
Stunt, alleged advertising, 271
Sturdee, Admiral, 205.
Submarines attack by, 255; contempt for, 252, 257
Success and failure, 184, 247, 289
Sun in Antarctic, return of, 73, 132, 209
Sunsets: in Atlantic, 35; at Grytviken, 277.
Sunshine, effect of, 222.
Swarthmoor Hall, 7
Swinburne, poetry of, 50, 62, 290.
Switzerland, visit to, 179
Sydenham, life at, 22-29
Sydney, lectures in, 112, 158, 248
Sympathy, 56, 100, 245, 278, 286.
Syndicate suit, 240

Tabard cigarettes, 92, 148
Tacoma, lecture at, 249.
Taft, President, reception by, 173.
Tagus, *Quest* in, 275.
Talcahuana, visit to, 41
Tank, military, referred to, 208; models, of 253.
Tantallon Castle, voyages on, 52.
Tariff Reform, 95
Telegrams, 155, 160, 231.
Telephone talks, 257, 260.
Temperament, hasty, 17, 80.
Tenderness, 173, 288.
Tennyson : poetry of, 21, 70, 290; quoted, 30, 57, 183, 213.
Tent in Antarctic, 67, 75, 135, 141, 144, 149, 197, 217.
Terra Nova · at Dundee, 83; as relief ship, 89; on Scott's expedition, 184, 187, 189.
Terror, H.M.S., 14, 15; Mount, 15, 128.
Tetrazzini singing : in honour, 160; in opposition, 173.

INDEX

Thames, sailing from, 45, 48, 110, 202, 273; Nautical College, boys from, 280.
Thanksgiving service, 156.
Theodolite, use of, 74.
Thirst, 222, 226
Thom, Capt, on *Southern Sky*, 229.
Thomson, Sir Wyville, 16.
Three Rock Mountain, 20.
Thunderstorm, 18, at sea, 41
Tidebrook, Sussex, staying at, 50, 60; revisited, 159.
Tierra del Fuego, 238.
Times, The, 195
Tintagel Castle, third officer on, 53
Titanic, loss of, 187
Tobacco business, 92, 190.
Tocapilla, adventure at, 41.
Torquay, *Nimrod* calls at, 111.
Towing: of *Nimrod*, 115; of *Terra Nova*, 83
Trade, development of British, 262
Train, special, to Dundee, 93; in Chile, 240.
Trans-Antarctic expedition Bruce's project, 184, 193; plans of Imperial, 194-196; failure of, 211, 213, 247.
Transport, polar, importance of, 132, 153, innovations in, 106
Treasure, hidden. romance of, 21, 26, 57, 217, 281, 285.
Trinity College, Dublin, 6, 9, 19
— House, membership of, 161, 188
Tripp, L. O H, on *Aurora* episode, 241, on *South*, 245-246.
Tromsö, visit to, 256.
Troops. American, in convoy, 255; Russian, transport, 97
Troopship, *Tintagel Castle* as, 53
Truant days at Dulwich, 26, 260.
Tula, brig, 13.
Tunnel to Antipodes, 20.
Tussaud's, effigy at Madame, 163.

Umtali, transport, 260.
"Uncle Shackleton," joke, 178.
Unconventionality, 86, 257
Undulations of Barrier, 138, 139.
Unexplored Antarctic, maps, 11, 196
Union-Castle Line, service in, 51.
Union Jack, presented by Queen to *Nimrod*, 110; at farthest south, 144; presented by King to *Endurance*, 202; return of, 249; presented by King to *Quest*, 273

United Methodist quoted, 50
Unknown, glamour of, 174, 288.
Uruguay government: assists, 232; is thanked, 239, 240; honours funeral, 279
Uruguay, Argentine cruiser, 83, 211, 232; Uruguayan cruiser, 279

Vahsel Bay, 208.
Valparaiso, visits to, 32, 42, 239, 240, 255
Vancouver, lecture at, 177.
Varanger Fjord, 257
Venesta boards, 110.
Verde, Cape, 55
Verses · by Richard Shackleton, 5, 19; by E. H S., 46, 47, 95, 275
Vestris, passage on, 253.
Victoria Land, 15, 46, 63, 127.
Victoria, Princess, 201.
— Queen, accepts copy of *O H M.S.*, 55
Victorian Order, 110, 162, 293.
Vienna lectures at, 170, 179; Geographical Society, 193
Vigilance on the floe, 219, 228
Vincent, J, on *James Caird*, 223.
Vince's cross, 110.
Virginian, passage on, 177.
Vivacity in argument, 56
Voice, peculiarities of, 56, 57, 84, 96.
Vostok, Russian ship, 12

Wade, A C, appreciation by, 178.
Wales, Prince of (King George v.), 161, 162
Walker, George, 92.
Wallstown, Cork, 9
Walworth, Browning Settlement at, 163, 199, 266.
War Napoleonic, 12, South African, 53; declaration of Great, 202; news of, 227, 230; services during, 248-264; end of, 263; spirit in America, 249.
— Office, 257
Washington, visits, 173, 249, 255.
Water, shortage of, 221, 234
Waves: in Weddell Sea, 220; of Southern Ocean, 225, of South Atlantic, 277.
Wealth. distribution of, 113; pursuit of, 191; prospects of, 111.

INDEX

Weddell : James, 12, 153 ; Quadrant, 206
Weddell Sea. Weddell in, 12, 153, Ross in, 15 ; Dundee whalers in, 39 ; Nordenskjöld in, 83 ; Bruce in, 88, Filchner, in, 189, proposed expeditions to, 184, 194, *Endurance* in, 206–212 ; drift on ice-floes in, 207–221
Wellington, N Z, visits to, 112, 158, 241, 244
Welsh Shire Line, service on, 45
Whales, Bay of, discovered, 117
White House, reception at, 173
— Island, visit to, 67, 68
— Sea, 257
Wild, Ernest, on *Aurora*, 242
— Frank on *Discovery*, 71, on *Nimrod*, 108, at Cape Royds, 125 ; depot-laying, 134, on Southern journey, 135, 139, 140, extracts from diary, 146, 147, sights Bluff depot, 149 ; gets medal, 162 ; at Queen Mary Land, 190, on *Endurance*, 199, 205 ; on the floe, 216, 219, 221, in charge at Elephant Island, 223, 235, at Spitsbergen, 256 ; as lecturer, 266 ; in Africa, 268 ; on *Quest*, 274, 278
— Cape, 222
Wilhelm II, Emperor, 170
Wilkes, Admiral Charles, U S N., 14, 173
Wilkie, Mr, Labour M P, 95
Wilkins, G, on *Quest*, 274
Wilkinson, Margaret (ancestor), 5
Wilson, Dr Edward A (" Bill "). on *Discovery*, 58, 68, 70 ; on Southern journey, 75 ; invited to join *Nimrod*, 106 ; death of, 189
Wind and sledge-travelling, 145, 148 ; and drift of floe, 215, in Southern Ocean, 225.

Winnipeg, lectures at, 176, 177.
Winter : on *Discovery*, 70, 71, at Cape Royds, 129, in *Endurance*, 208 ; equipment for, 256, invasion planned, 250 ; at Murmansk, 261.
Wireless at Falklands, 231 ; on *Quest*, 272, 278
Wit, Irish, 290
Women · sympathy of, 198 ; votes for, 96.
Woodville, takes body to South Georgia, 279
Woosnam, Rev G. W, 27, 30, 33
Wordie, J. M, on *Endurance*, 205 ; appreciation by, 281.
Wordsworth quoted, 17
Working men, sympathy with, 95 163, 199, 288
Worries in business, 186
Worsley, Capt Frank A · in *Endurance*, 199, 205 ; in Weddell Sea, 221, 223, in the *James Caird*, 223, 225, 226 ; crossing South Georgia, 227 ; on Elephant Island relief ships, 231–235 ; left behind, 242 ; on the *Ella*, 257, at Murmansk, 258, on the *Quest*, 274
Writ served in Ottawa, 174
Writing, dislike of, 260.

Yacht Club Royal Clyde, 201 ; 242 ; Royal Harwich, 59
— Squadron, Royal, 272
Yarmouth, I.W, 60.
Yelcho and Elephant Island relief, 233, 235, 237, 238, reception at Valparaiso, 239
Yorkshire caution, 154, 290 ; origin of family, 4
Young, Douglas, 232
— Sir Allen, 165
Younghusband, Sir Francis, 266

Zurich, lecture at, 179

PRINTED BY MORRISON AND GIBB LIMITED, EDINBURGH

Ingram Content Group UK Ltd.
Milton Keynes UK
UKHW021809030723
424469UK00008B/438